Laboratory Information
Management Systems

Laboratory Information Management Systems

Second Edition, Revised and Expanded

Christine Paszko
Accelerated Technology Laboratories
West End, North Carolina

Elizabeth Turner
USACE Washington Aqueduct
Washington, D.C.

CRC Press
Taylor & Francis Group
Boca Raton London New York

CRC Press is an imprint of the
Taylor & Francis Group, an **informa** business

Preface

The concept of laboratory data management is not a new one. The evolution in *Laboratory Information Management Systems* (LIMS) over the past 20 years is astounding. There have been many changes in technology, including but not limited to hardware advances, improved software development tools, communications, and networking, that have accelerated the development of LIMS. It is also important that the reader be aware that a LIMS is more than software; it is a process of integration that encompasses laboratory workflow combined with user input, data collection (instrument integration), data analysis, user notification and delivery of information and reporting. This book attempts to provide the reader with an easy-to-understand explanation of each aspect of a LIMS and a better understanding of what is important in selecting and implementing a successful LIMS project. LIMS have evolved from notebooks to spreadsheets to simple laboratory data acquisition and storage systems to complex relational database systems that integrate laboratory information with enterprise-wide computing environments to facilitate rapid access of information throughout an entire organization. The goal of this book is to introduce the concept of a LIMS, LIMS features, an examination of the underlying technology, and a look at the human factors involved.

To assess the future direction of LIMS, it is necessary to understand how the current systems have evolved. Chapter 1 provides a brief historical perspective on the evolution of LIMS and the technology that has enabled the data management revolution, and introduces future technological trends that will fuel the development of LIMS. Commercial LIMS have been around since the 1980s. In addition, many organizations have designed and implemented an in-house or "home brew" LIMS. The organizations that utilize a LIMS vary greatly, from research laboratories to manufacturing laboratories to commercial testing laboratories; however, they are all basically organizing their information to make quicker, more informed decisions and to share that information.

The next portion of the book deals with the human element, the laboratory processes, and the people. Far too often this critical piece of a LIMS implementation is overlooked. The LIMS planning committee can spend considerable time and resources in selecting a LIMS. However, if those in the laboratory are not allowed to participate in the process and are simply led to the LIMS, the results may be less than favorable. The input from laboratory personnel in LIMS selection is vital to selecting the right LIMS for the laboratory. One very real issue that hasn't been addressed in the past is the psychology of implementing a LIMS. Obtaining group acceptance of change in the laboratory infrastructure and incorporation of the needs of specific groups is critical for successful implementation. Additionally, many fear that automation (LIMS) will eliminate or seriously threaten their position in the company. These fears are real and need to be addressed.

The next section examines LIMS in more detail and provides an overview of how they are used across industries. Although the functions and requirements of a LIMS may vary by industry, there are shared concepts fundamental to all LIMS. We review a LIMS framework and discuss what the fundamental elements of a LIMS should include, regardless of laboratory type. The features available in a LIMS are numerous and growing rapidly. Some features are basic to all LIMS while others are industry-specific. The role of the Internet as it relates to LIMS is also explored. Specific LIMS features integral to many LIMS, such as audit trails, automatic reporting capabilities (via e-mail, fax, or printed copy), import/export capabilities and data warehousing will also be discussed.

The remaining chapters deal with more technical issues, including establishing laboratory requirements, outlining/ranking desirable features in a LIMS, database design considerations, hardware and operating system requirements, and other fundamental considerations in selecting a LIMS. Since many readers may have the responsibility for preparing a request for proposal (RFP), a chapter has been included that outlines critical elements in preparing a solid RFP and provides an example. Following the RFP process, evaluation criteria must be established so that when the proposals are received there is already a mechanism in place to evaluate the responses. The process of selecting and implementing a LIMS can be relatively painless if realistic selection criteria and a practical implementation plan are developed and followed. In addition to satisfying basic laboratory functions such as sample tracking and data entry, a LIMS must also comply with regulatory requirements and electronic data security in a laboratory. Regulations are often industry-specific. An entire chapter is dedicated to reviewing the regulatory requirements with which a LIMS must comply for various industries. Hardware and operating system requirements of a LIMS must also be carefully considered. Software and hardware compatibility, network design, and resource utilization are critical to the optimization of a LIMS.

A LIMS implementation plan should be developed once a LIMS has been selected. Chapter 11 discusses key considerations, compares phased versus "shotgun" implementation approaches, and provides a sample implementation plan. Validation is the last step of the LIMS implementation and often a continual step as new functionality is incorporated. The validation process ensures that the LIMS will meet specifications and conform to predefined quality assurance criteria. Requirements of a successful LIMS validation are discussed.

The final chapters cover the development of implementation and validation plans of the LIMS. It is important to realize that purchasing a LIMS is not a static process; rather, it is constantly changing in response to the laboratory data management requirements. Advances in technology occur, these advances are transferred to the laboratory environment. One example is, Web access to laboratory results. We have provided several pages of additional LIMS resources, including a list of LIMS vendors, a listing of Web sites, a comprehensive glossary of terms, and a suggested reading list. The authors wish to thank Don Kolva and Lisa Gorenflo of ATL, Inc. for their critical review and encouragement; Tom Jacobus and Lloyd Stowe of the USACE Washington Aqueduct; and Kevin Dixon of the NJ-American Water Company for their support. We also wish to thank Annie Cok of Marcel Dekker, Inc. for her support and assistance; and David Turner, who put up with Elizabeth usurping their computer for numerous months, for his support and encouragement. It is our hope that after having read this book the reader will have a solid LIMS knowledge base on which to build.

Christine Paszko
Elizabeth Turner

Contents

Preface   *iii*

1. Historical Perspective   1

2. LIMS Fundamentals: Overview of Laboratory Information
    System Development and Project Planning   7

3. Data Management and Basic LIMS: Functional Requirements
    and Features   23

4. Data Management and Advanced LIMS: Functional
    Requirements and Features   35

5. Life Cycle of LIMS Software Development   47

6. Regulatory Requirements   55

7. Hardware and Operating System Requirements   69

8. Obtaining Laboratory Personnel Input   83

9. Critical Elements in Preparing a Request for Proposal   93

10. LIMS Evaluations   101

11. Enhancing Data Quality with LIMS   107

12.   LIMS Validation   113

Appendix A: Sample Request for Proposal   *163*
Appendix B: Sample Scripted Demonstration   *207*
Index   *225*

1
Historical Perspective

Before we can discuss Laboratory Information Management Systems (LIMS), we must first understand the technology and tools that enabled the creation of these sophisticated software packages that are replacing scientists' notebooks. Table 1 outlines a brief description of the technological events relevant to this through today. With the pace that technology is moving, there will undoubtedly be many more advancements after the publication of this book. The evolution of LIMS is an interesting one. In the beginning, there were scientists with laboratory notebooks and everything was hand-written: dates, experimental designs, results, comments, observations, and more. In the early days of computer-based LIMS, there were host-based systems connected to terminals by serial lines: all of the processing was performed on the host. Upon completion of the processing, the host would post results to the terminals within the laboratory. These systems lacked flexibility.

Only very wealthy companies had access to these early systems and advanced technology. They made a commitment to the LIMS concept despite the high cost because they realized that those who could deliver the correct information, ahead of their competition and at a competitive price, would emerge as the market leader. These industries understood that knowledge truly is power. The same is true today. Only today, in addition to faster and better, the market also demands even more affordable information management solutions. Today client/server LIMS architecture as shown in Figure 1 is increasingly popular.

Table 1   A Brief Chronology of Computers

	5000 years ago, it was the abacus; that was eventually replaced by paper and pencils.
1623	German scientist Wilhelm Schikard invents a machine that uses 11 complete and 6 incomplete sprocketed wheels that could add and, with the aid of logarithm tables, multiply and divide.
1642	French philosopher, mathematician, and physicist Blaise Pascal invents a machine (the Pascaline) that added and subtracted, automatically carrying and borrowing digits from column to column.
1694	Gottfried Wilhem von Leibniz (1646–1716) improves the Pascaline by creating a machine that could also multiply. Like its predecessor, Leibniz's mechanical multiplier worked by a system of gears and dials.
1822	The Difference Engine is designed by British mathematician and scientist Charles Babbage. Babbage's assistant, Augusta Ada King, the Countess of Lovelace (1815–1842) and daughter of English poet Lord Byron, is instrumental in the machine's design. In the 1980s, the US Department of Defense named a programming language ADA after her.
1930	American electrical engineer Vann Bush produces the first partially electronic computer called a differential analyzer, capable of solving differential equations.
1942	American theoretical physicist John V. Atanasoff and his assistant Clifford Barry build the first computer that successfully uses vacuum tubes to perform calculations. The machine is called the Atanasoff Berry Computer, or ABC.
1944	At Harvard University, the Harvard–IBM Automatic Sequence Controlled Calculator is developed under the direction of Howard Hathaway Aiken. It contained more than 750,000 parts and takes a few seconds to complete simple arithmetic calculations.
1945	At the Institute for Advanced Study in Princeton, Hungarian–American mathematician John Von Neumann develops one of the first computers used to solve problems in mathematics, meteorology, economics, and hydrodynamics. Von Neumann's Electronic Discrete Variable Computer (EDVAC) is the first electronic computer to use a program stored entirely within its memory.
1946	The first automatic electronic digital computer, ENIAC, constructed at Harvard University by electrical engineers John Presper Eckert and John William Manchly in consultation with John Atanasoff. The electronic numerical integrator and computer contains radio tubes and runs by electrical power to perform hundreds of computations per second.
1946	The word "automation" is used for the first time, by Ford Motor Company engineer Delmar Harder to describe the 14 min process by which Ford engines are produced.

Table 1 Continued

1948	At Bell Telephone Laboratories, American physicists Walter Houser Brattain, John Bardeen, and William Bradford Shockley develop the transistor: a device that can act as an electrical switch.
1949	British biochemist Dorothy Crowfoot Hodgkin is the first to enlist the aid of an electronic computer in discovering the structure of an organic compound: penicillin.
1951	The Univac computer is introduced for business use by Remington Rand.
1953	IBM introduces the IBM 701, the first computer for scientific and business use.
1955	The IBM 752, the company's first computer designed exclusively for business use, is produced.
1959	The microchip, an integrated circuit made of a single silicon wafer, is invented by American engineers Jack Kilby of Texas Instruments and Robert Noyce of Fairchild Semiconductors.
1960s	Raymond Goertz at Argonne National Laboratory in Argonne, Illinois, and Ivan Sutherland at the Massachusetts Institute of Technology in Cambridge, Massachusetts, demonstrated early versions of head-mounted displays (HMDs) used in virtual reality.
1960s	Computers come into common usage in government and industry, but for many years they are not available to most consumers.
1970	U.S. scientist Ted Hoff, working for Intel, invents the microprocessor, a silicon chip containing the central processor of a computer. The versatile chip will lead to the proliferation of small inexpensive computers for home and business use. Intel microprocessors will be marketed commercially in 1971 for the first time.
1973	The Internet is created in large part by American computer scientist Vinton Cerf, as part of the United States Department of Defense Advanced Research Projects Agency (DARPA).
1974	A text-editing computer with a cathode-ray tube video screen and its own printer is put on the American market by Vydek.
1975	The first personal computer, the Altair, is put on the market by American inventor Ed Roberts. The Altair 8800 uses an 8-bit Intel 8080 microprocessor, has 256 bytes of RAM, receives input through switches on the front panel, and displays output on rows of light-emitting diodes (LEDs).
1975	Americans William Henry Gates III and Paul Gardner Allen found Microsoft, which will become the world's most successful manufacturer of computer software.
1976	Ironically just a year later, Tagamet, a drug for the treatment of ulcers, becomes available. By 1990, it will be the most frequently prescribed drug in the United States.

Table 1 Continued

1977	The Apple II computer is marketed by American inventors Stephen Wozniak and Steve Jobs: the first personal computer to be accessible not just to hobbyists but to the public at large.
1980	The first IBM personal computer, employing the Microsoft operating systems MS-DOS, is marketed, with great success. The 1980s bring small, powerful, and inexpensive computers to American households. Each new innovation in computer hardware encourages better software, which in turn encourages the production of better computers.
1984	The first 1 megabyte random access memory (RAM) chip is developed in the United States by Bell Laboratories. It stores four times as much data as any chip to date.
1984	The development of Internet technology is turned over to private, government, and scientific agencies.
1990s	Computer use continues to proliferate as hardware costs continue to decline and the Internet and software development continue to grow.
1995	Global positioning system (GPS; Navstar) is declared fully operational, total development costs reach 10 billion.
1996	Intel introduces the Pentium 200 MHz processor and Microsoft ships Windows 95.
1997	Intel introduces a new 2-bit flash memory that validates Moore's Law, which states that each new generation of chip will be capable of processing twice the capacity of its predecessor.
1998	Launch of DVD-ROM drive with 5.2 gigabyte (GB) re-writable capacity. Apple computer launches iMac.
1999	Intel Pentium III processor is introduced with unique ID embedded for web identification. IBM announces 73.4 GB drive.
2000	Windows 2000 launched by Microsoft. Intel ships 1 GHz processor. United States District Judge Thomas Penfield Jackson orders the breakup of Microsoft into two companies: one producing operating systems and the other producing applications programs after a landmark case in which Microsoft fails to prove that its business practices were not in violation of antitrust laws.

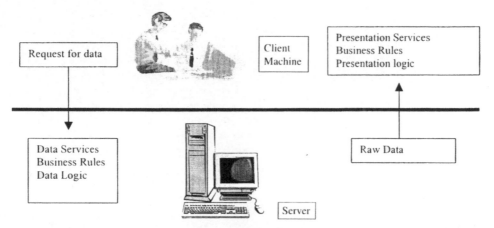

Request for data

Client
Machine

Presentation Services
Business Rules
Presentation logic

Data Services
Business Rules
Data Logic

Raw Data

Server

Fig. 1   Schematic diagram of client/server architecture. SQL server is an example of a client/server database management system. This means that the client machine processes, which are often running on remote machines, must communicate with the SQL server over the network as in the figure. ODBC is a native SQL Server driver used by many applications and developer tools, including Visual Basic and C++.

2
LIMS Fundamentals
Overview of Laboratory Information System Development and Project Planning

A solid design is critical to any Laboratory Information Management System (LIMS) or other software development venture. Most commercial LIMS vendors have a sound understanding of good database design, although not all have. For consumers, it is important to perform their own needs assessment and then determine which LIMS meets their specific requirements.

I. BUYING OR BUILDING

Although some laboratories decide to create their own systems, this often represents an expensive venture for many reasons and few repeat the effort. Many programmers who lack database design expertise are overzealous when they provide estimates for laboratory managers or owners. For many it is their first attempt at creating a LIMS and there is no historical experience to draw from. Projects typically take much longer than anticipated to complete, the laboratory end-users must perform all the software debugging (which is costly), and there is often poor documentation and core personnel turnover.

In today's rapidly changing technology environment, it may be wiser to purchase an off-the-shelf LIMS and utilize the laboratory's programmers to enhance it with add-ons and custom reports unique to the laboratory. There are many well-written commercial systems on the market for a variety of laboratory types. Major advantages of commercial systems are that technical support and a team of experienced software engineers stand behind the product and have the financial incentive to keep their product current, migrating it to

7

the latest platforms, adding functionality, and providing assistance when your laboratory needs it. Most successful businesses adhere to their core competencies and recognize when it makes business sense to outsource. In any case, the same factors that are required in the inhouse development of a LIMS should be considered in the selection process of a system from an outside vendor. Before the decision is made to purchase a LIMS, the laboratory should have a good understanding of its operations, how it wishes to improve those functions, and how the software will help it attain those goals. We review here the development process through implementation, and provide readers with some important factors to consider before undertaking these tasks.

II. PROCESS OF CHOOSING AND IMPLEMENTING

The process of choosing and implementing an LIMS involves six steps or phases as outlined in the American Society for Testing and Materials' (ASTM's) guidelines and listed below. The process is diagrammed in Figure 1.

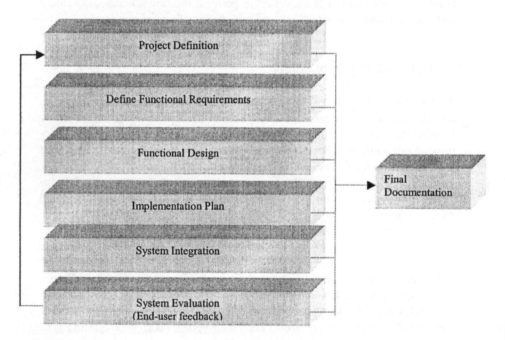

Fig. 1   Standard steps in development of a software system

Project Definition
Functional Requirements
Functional Design
Implementation Design (implementation, training, operation and support)
System Integration
System Evaluation

A. Project Definition

Project definition consists of a short description of what is to be achieved, by whom, when, and why. This document is typically one page long. If the project is complex and multiple pages are required, an executive summary should be included. The project definition is critical for the entire project and should be thought of as the foundation. Once into the project, it is very difficult to change the direction or foundation of the project, therefore this phase should be developed very carefully. It is characterized by feasibility determinations, deciding on an off-the-shelf product or inhouse development, developing and documenting the project definition.

In this planning phase of the project, all aspects are considered, including integration with other systems (such as Enterprise Resource Planning [ERP], financial, etc.), instrument integration for automatic data acquisition, and long-term maintenance and growth (future data migration). Vendors of ERP are numerous: SAP, PeopleSoft, J.D. Edwards, and ORACLE, among others. Key features of ERP include global financial capabilities, advanced planning and scheduling, product configurators, supply chain management, customer relationship management, e-commerce, business intelligence, and component (object-oriented) architecture. Many large companies see the advantages of an ERP system (supply chain management) which include a better understanding of costs and inventories, as well as the ability to react to competitive pressures, accelerate production, and better understand financial closing cycles. In addition to these benefits, companies benefit from them in attempting to globalize their business, improve customer service, improve the availability of information, and web-enable their business. An ERP system can also allow companies to reduce their costs and improve productivity, standardize business, integrate and improve business processes, increase flexibility, and integrate acquisitions.

This phase is characterized by the gathering of several groups or teams to understand their needs and agree on the required features for the successful project outcome. It is important to hear the needs of all parties that will be involved in the project to ensure that the final product will fit the organization and not just a few subgroups. Once the document is complete, it is important to have everyone sign off on the original document as well as any changes made to it.

B. Functional Requirements

The functional requirements step involves all user entry requirements and system output requirements being described in detail. Also described in this phase are any integration functions required to produce the outputs from the user entry requirements. An example is the need for analytical instrumentation in the laboratory to be interfaced with the LIMS for final result outputting. Since this phase is quite comprehensive, it is often a good idea to break the project into several phases. In that way the entire task from definition to implementation does not seem as daunting. This documentation should be sufficiently detailed to allow software engineers to develop a functional design of the database or to select a commercially available system. Additional information to be included in this document includes the project's objectives, resource requirements (financial and human), and system specifications. The functional requirements document serves as the request for proposal sent to potential bidders, minus the budget information and any proprietary information. Information that should be included in the functional requirements includes:

- Overview of the system: Context and constraints of the system.
- Objective: State the objective of the project (or portion of the total project) covered by this functional requirements document. This must be consistent with the project definition.
- Specific goals and expected benefits: A detailed list of objectives and expected benefits of computerization. Assign priorities to each item on a scale from 1 (lowest) to 5 (highest). It is important to be specific so that this list can be used for subsequent evaluation.
  Nature of the project: A description of the process, procedure, test, experiment, function, or operation to be computerized. Include background references or examples when possible. Describe any alternative methods that may be used to produce the desired result.
- Describe each piece of instrumentation and or other software packages to be integrated to the LIMS in the project. List the instrument or software name and version, model, and manufacturer. In cases where the LIMS must handle data acquisition, the following additional information is required: a brief description of the equipment, functions, and communication standards.

Additional information that is helpful includes formats (output files should be included where possible) and protocols that are utilized by the equipment, equipment location, required fields for integration, and information relating to networking of the instruments. It is often helpful to provide the LIMS vendor with an organizational chart as well as a schematic diagram of the building including where each computer and instrument is to be located.

C. Functional Design

This is the phase in which detailed documentation is produced to describe the system and detail how the functional requirements are to be achieved. It is independent of the hardware and software requirements of the LIMS and characterized by flow diagrams of the entire process (information flow throughout the laboratory and beyond), implementation diagrams, and Gantt charts.

D. Implementation Design

The hardware and software are selected next. Their selection often depends on many factors: best available technology, budget, current infrastructure, existing hardware, expertise of the information technology (IT) staff, license fees, and other factors. Procedures for rolling out the new system, training for end-users and the database administrator, and continuing support are included in this phase. This document may include any alternatives in the implementation designs that can be evaluated by the system integrators. Sometimes inconsistencies are uncovered among functional requirements or goals of different groups during the implementation design process. They are typically resolved within the group. The implementation design document should be complete enough to allow straightforward implementation, but not so buried in details as to lose sight of the "big picture."

E. System Integration

The system integration phase consists of putting all the pieces of the system together. This includes gathering all the required components, interfacing the system components, installing software, and, finally, "going live."

F. System Evaluation

In this final phase the project definition and functional requirements are revisited and compared to the final installed system to determine how well the requirements were met. Users often utilize a checklist to compare before and after the installation of the system to ensure that all their needs have been met. If some requirements were not accounted for or were missed, these issues are addressed in the cycle just described. It is also important to build in an ability to grow with the system and not have this be an afterthought. Consider the changing needs of the organization, identify future features that may be needed in the system, and be sure that an upgrade or migration path exists to accommodate them. Although it is not easy to predict what the laboratory will require in the future, it is important that the LIMS software be adaptable and able to accommodate change. End-users can determine this by asking how easily changes to the screens/reports can be made, if a migration path for the product exists, and by avoiding proprietary database systems. Once you have asked these questions, the answer should be "Yes" to all of them. By considering growth and long-term planning, you may reduce the cost of obsolescence and may also want to investigate alternatives to purchasing, such as leasing. The lessons learned from going through the process once will help in preparing for system improvements, enhancements, upgrades or installation of other systems.

System documentation is not a separate phase but is prepared as the project proceeds. It consists of project definition documents, functional requirements, functional design, implementation design, system integration, final phase, and system evaluation documentation. Additional documentation includes user manuals, training manuals, and enhanced documentation for the database administrator. Coordinated documentation is critical to the success of the project, since it is final record of the entire process. If there is employee turnover, new employees can use the documentation to understand and learn the process.

The ASTM recommended process has been updated and modified and is diagrammed in Figure 1. The entire process occurs as separate steps in Figure 1. However, there is actually tight integration between the different steps that requires cooperation between the many different people involved in the project, including IT staff, specialists in database design, managers, implementers, and end-users. The input from different groups is required at different stages of the process. This makes software development projects a challenge, since each group has its own objectives, agendas and ideas, and, unfortunately, its own terminology. Integral to any software project is understanding the other groups' (IT, quality assurance (QA), management, etc . . .) terminology. Once all members in the group understand the others' goals and objectives, projects can move forward. Because there are often many different groups involved in an LIMS project, conflicts over functional requirements may arise. It is important to address these conflicts to ensure that the entire group's needs will be met by the LIMS. Fortunately, with the graphical user interface GUI and Windows technology, most LIMS are flexible enough to allow users to modify screens, add modules, and turn functionality on and off for certain users and groups of users. This flexibility has allowed LIMS users to overcome many of the conflicts of the past. When conflicts do arise, the LIMS project leader's diplomacy can often find a way to satisfy all parties. Some may disagree with this approach, but since all users have a stake in the implementation, it is important that the functional requirements of the various groups are met. Otherwise, one group may not utilize the system and there will be a gap in the "data picture."

For the purpose of this book, and for most projects two design phases are adequate: functional design and implementation design. The first is hardware- and software-independent; the latter is not. While the system or customization to the system is under construction, it is vital to adhere to the specifications established early in the project. If there are problems in following the original project specifications, they must be revisited, modified, and implemented. It is desirable to have the key member of the project team, including someone from the QA/quality control (QC) group, involved in this testing and release phase.

According to the ASTM, complex projects or large-scale systems often require multiple phases. These may include organizational design, system user job design, task modeling, system architecture design, software architecture design, system user interface design, data modeling and database design, software pseudocoding, electrical and mechanical interface design, and electrical and mechanical component design. However, since there are many good commercially available LIMS available, much of the design will already have been established, tested, and refined. This saves the end-users considerable time and effort and keeps them from being forced to reinvent the wheel.

III. SYSTEM VALIDATION

Quality assurance and quality control steps should be integral to any software project and are central to each step in the development cycle. System validation often refers to software systems used in regulatory environments. The primary purpose of system validation is to ensure that the software is performing in the manner in which it was designed to. For example, the system acceptance criteria should be established and tested against quantifiable tasks to determine if the desired outcome has been achieved. Task items (features) must be quantifiable, such as autoreporting, reproducibility, throughput, or accuracy. This check ensures that the entire system has been properly tested, incorporates required controls, maintains, and will protect data integrity. A method must also be established to handle the validation process and associated documentation. How are anomalies ("bugs") documented and repaired? How are modifications to features or calculations handled that are incorporated following initial validation? Although vendors perform internal system validation, once modifications and customization are added by the end-user, vendor, or a third party, the system must be revalidated to cover and incorporate the changes. In addition, different groups using the system require their own validation based on their specific requirements. These requirements must be outlined in the initial phases of the development cycle, typically in the functional requirements phase. Different types of laboratories have different regulatory requirements, some of which are discussed in Chapter 6. The vendor, internal QA staff, or third-party consultants can perform system validation. Their documentation becomes part of the entire system documentation.

IV. FEASIBILITY CONSIDERATIONS

Before a project can proceed, feasibility must be discussed. Questions to be answered include whether project resources (human and financial) and in-house expertise (IT, project management, etc) are available; anticipated advantages; software and automation to be used. Additional considerations include anticipated changes in the market that may pressure the organization to remain competitive, integration of the system into the organization, and compatibility with other enterprise systems being used. It is important to determine the problem that software or automation is supposed to solve. This requires a clear and realistic understanding of the current operation, resources, and inhouse expertise and will serve as a guide to what resources need to be obtained not only to put the system into place but also to maintain and upgrade it into the future. Since these projects typically involve a significant amount of resources, it is import to consider many factors when creating the project definition document. Table 1 lists some of these factors.

It is important to describe how current operations will improve following automation, visualize future operations, and then work toward that description. Before the project can move forward, a careful cost/benefit analysis must be performed. This analysis determines whether the investment will result in significant financial gain to justify the project.

Table 1 Sample Considerations in Resource Planning

Automation Considerations	Current Situation	Planned
Work volume and flow		
Cost of operations		
Inhouse staff computer (IT) expertise		
Equipment available		
Equipment required		
Resources (human and financial)		
Space considerations		

To determine this, project managers must consider alternatives to automation and the expected outcome of each alternative. One alternative is always to do nothing, but the fact that computerization and automation are even being considered means that there is an existing problem managers are trying to solve. A second alternative is to refine existing methodologies and process flow. This is not always easy, since most organizations have already optimized their process to increase productivity. However, if the organization has not already optimized operations for efficiency and productivity, some of the areas to examine include improving manual systems, reducing steps, adding personnel, adding equipment, and providing efficiency training. Also to be considered might be more involvement by management, acquisition of commercially available software for certain parts of the operation, and custom system development (with a vendor or internally). Once a list of viable alternatives is constructed, project managers need to estimate the total cost of each. Cost estimates should include internal and external resources required to achieve each alternative. Estimates should be typically within the range of 15–20% accuracy. Project managers may want to consult with the organization's financial advisors, who may have more experience in estimating total cost for each alternative. Once the cost/benefit analysis is complete, consider the risk of each alternative: not only the potential risks of proceeding with each, but also the risk of not proceeding or the risk of doing nothing.

Once a decision is made to proceed, and a particular alternative chosen, regroup with the project team because someone's alternative of choice is probably not the one favored by the financial analysis. Once all integral participants have been identified (typically upper management, financial management, information technology, end-users, supervisors), it is important to negotiate a project definition agreement to encompass all the required functions. The pre-

Table 2 Elements of a Project Definition

Documentation of the current situation and the planned improvements following implementation.

What is the current problem or area that improvement is required?

What is to be gained by the successful implementation of this project?

What benefits are expected from the outcome of this project?

What benefits are expected from this project than can be measured? Productivity? Turnaround?

Results of the cost–benefit analysis and project justifications.

What are the estimated project costs?

What are the project goals?

Identify any constraints that may interfere with the project completion

Identify project limits.

Identify and define risks for each alternative.

Prepare completed project definition document that has been signed-off by development team.

Prepare a Gantt chart for entire project and regularly update it.

---

vious definition document may not require alteration or slight modifications may be required to satisfy particular groups.

Once the alternative has been selected and the project team concurs on the definition document, each group should list expected benefits to their group, including goals that the system implementation can be compared against. An example might be a reduction in transcription errors in automatic data migration from instrumentation and increased efficiency, resulting in increased capacity for employees. Each team member should also document what his or her involvement in the LIMS project will be (i.e., what they can do or provide to assist in the implementation) to ensure success. Remember that software does not succeed: People do. LIMS is a team effort. It is important that the project definition document be succinct, but it should cover the items listed in Table 2.

Origins and types of input data include the following: data input into the system from interfaced instrumentation, via another software package or another source, end-user inputs, automatically scheduled events, user-definable upper and lower limits, significant figures, and required units.

A. Automatic Calculations and Output

One of the functions that computers perform best is routine calculations, so it is only natural to transfer routine calculations to the LIMS. These items are typically easily validated. When these functions are transferred, it is important to document what the initial raw data represents, the actual formula for the calculation, how it is executed, and the availability of these raw data to the operator and others in the laboratory prior to data manipulation.

B. Reporting Requirements and Final Output

One of the greatest advantages of automation is the ability to deliver "real-time" information to those who make decisions. Prepare a list of the required reports and the elements included each report: which data; what format; location of information; delivery method (printed copy, e-mail, fax, voice), and to whom reports should be directed (often more than one person requires the information). Additional information consists of automatic or manual reporting, documentation on data location, length of time raw data are available, and how frequently it should be archived. Also included in the report should be user-definable reporting times, how to modify those times, and number of previous results.

V. OPERATIONAL CONSIDERATIONS

It is important to have a back-up plan in the event that system availability and reliability are hindered. Many servers today have redundancy built in. If possible, the system should be taken down for upgrades and major maintenance during off hours. There should also be contingency plans for backup hardware, in the event that hardware is damaged. Hardware should be on an uninterrupted power supply (UPS) to clean the electricity and to provide proper shutdown of the server and computers. A solid data backup procedure should be in place as well as routine backup rotation. For the latter, a copy of the data should be held off-site in the event of a fire or other mishap. In addition, only those that have authorization to the system should access the system. It is the database administrator's duty to ensure adequate network and system security, including proper use of user names and passwords, firewalls, and other necessary security measures.

Figure 2 reviews several different implementation paths. Note that implementation design is only part of the implementation process. Using information from the project definition, functional requirements, and functional design, the LIMS team should perform a market search to determine if the required system is available commercially or if vendors are willing to design

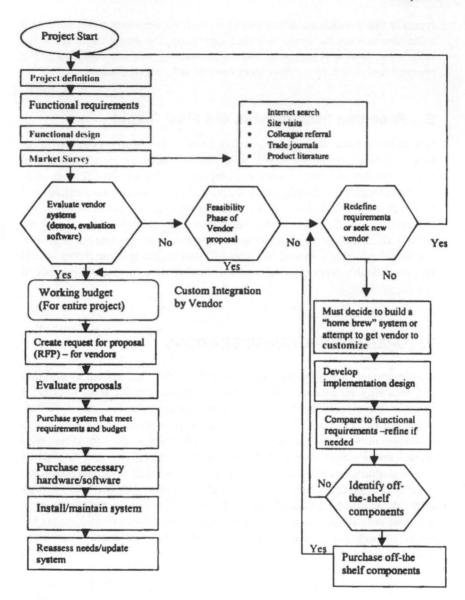

Fig. 2 LIMS implementation process

a custom system to match their unique data management requirements. If the latter applies, the flow is straightforward through to procurement, operation, installation, and implementation. If not, the LIMS team must decide whether to modify the system definition, to modify a commercially available system, or to build the system partially or totally inhouse.

The right-hand loop in Figure 1 documents inhouse implementation of the system. That loop calls for the implementation design to be developed, using the remainder of this section. At each pass through the loop, there is a choice between vendor-supplied or inhouse-built components. In each case, the system must be specified to the necessary level. The implementation design loop aids users in identifying certain elements in the system that cannot easily be changed or modified.

It is important to identify those elements in the system that cannot be changed. This may have an impact on hardware and software requirements for data output and manipulation. From there an acceptable system configuration can be proposed, with a suitable implementation design. Do, however, avoid proprietary systems at all costs and utilize standard components whenever possible. Review the proposed system configuration with the LIMS team and vendor or inhouse group to ensure that the major requirements are addressed, considering budget (final cost), expertise, and feasibility. If, during this review, new requirements are uncovered, it is important to revise the existing documentation, obtain approval from all of the original members, and to obtain signatures (sign-off) on the revised document. Different organizations have different procedures for budget approval, sending out requests for proposals, and for final procurement.

VI. INITIAL IMPLEMENTATION DESIGN: DESIGN POSSIBILITIES

Review the hardware and software requirements and select systems that will best meet your needs. Do not cut corners with hardware. With hardware costs constantly declining, laboratories can purchase very powerful servers at a fraction of what they cost 5 years ago. Consult your LIMS vendor and solicit suggestions. They typically have extensive expertise with both hardware and software platforms and are familiar with the advantages and disadvantages of each. Once you have determined the required hardware and software technical requirements, you will find that a number of potential implementation designs could work for your laboratory. There will undoubtedly be trade-offs, but ultimately it will depend on what makes sense for your laboratory, financially and with your particular inhouse priorities, LIMS requirements, and IT expertise.

It is also important to suggest several alternatives for implementation of the LIMS. These should be examined against the functional requirements, timetable, and availability of resources.

A financial analyst should be included in the team to address concerns about the return on investment (ROI). The IT group can compare the various hardware, network, and software configurations possible against the functional requirements of various laboratory groups. Automation efforts, such as instrument integration, integration with other software packages, and scheduling software, should also be analyzed in this manner. It is also important to consider the system's long-term maintenance requirements in the ROI calculations. Most systems today employ some type of client server configuration of their LIMS. The LIMS team needs to decide on the level of involvement they can realistically have in the LIMS project, which will be greatest during the implementation. If inhouse resources are limited, then the team may recommend purchase of a total system with support (turn-key system). This approach may be considerably more costly, since the vendor is responsible for the implementation. However, in the long run it may save the laboratory money, since the vendor is more experienced than the inhouse staff. This is not to say that the laboratory personnel will not be involved at all; they are critical to the success of the implementation. Another alternative is to purchase an on off-the-shelf hardware/software package and have the laboratory staff perform the implementation. This is especially applicable if the laboratory has the available expertise, resources, and experience from a previous LIMS implementation. If the laboratory lacks the required expertise, this method may take longer to achieve full implementation but the advantage is that the laboratory can go at its own pace. The most labor- and cost-intensive approach would be for the laboratory staff to purchase hardware components, customize the hardware internally, create their own software and documentation, and provide support.

The LIMS team must understand the benefits and limitations of currently available applications. This can be accomplished by several means, primarily by obtaining demonstration versions of the software, on-site scripted demonstrations, visiting other user sites, and reviewing the product documentation and web sites. A current trend among laboratories is to find a system that best matches their business needs, purchase the software in components or modules, and then allocate time and resources for customization and modification. This allows the laboratory staff to incorporate functionality for requirements specific to their laboratory. This can typically be accomplished by the vendor, with the internal IT group, or via an outside contractor.

VII. CONSIDERATIONS IN VENDOR SELECTION

Many searches for LIMS vendors begin today with the Internet. The initial search compares the functionality of system components/modules or entire systems with laboratory system requirements: experience of the LIMS vendor: vendors' abilities to integrate/interface instrumentation from different vendors into the LIMS; and a review of the types of support, implementation, and training services available. Other factors to consider when selecting a vendor include the availability of software, on-site or remote training, their certifications, documentation supplied with the LIMS, frequency of product upgrades, mechanism for fixing software anomalies, warranty, vendor's reputation in the industry and in specific application areas, any previous experience with the supplier, and feedback from other end-users and user groups.

A. Vendor Review

Once a vendor has been selected it is a good idea to review the initial project definition, general system requirements, functional requirements, and functional design with the vendor, paying special attention to the following items:

- Validity of the final, detailed proposal: Has anything in the laboratory changed?
- "Fit" of the proposed system to the specific requirements of the laboratory.
- Review of the Gantt chart: Is the timeline realistic? Are there any potential pitfalls?
- Review of the cost proposal, contract, license agreements, and implementation plan. Perform a final review with the vendor so that everyone has a clear understanding of the goals of the project and to ensure that it progresses smoothly.

B. Other issues to consider

These issues may include capability for future system upgrading; establishing liaison between suppliers and purchasers to coordinate specifications, scheduling, modifications, and acceptance testing; determining the existence and availability of adequate system and component documentation; finalizing delivery schedules; maintenance agreements or contracts.

C. Inhouse Implementation Considerations

The required people must be available for a sufficient amount of time to complete the project on schedule. Consider project leadership and responsibility, existing skills, learning curve effects, any required training and availability of specialists.

D. Availability of Resources

These may include the need for equipment configuration, documentation, and maintenance or troubleshooting.

E. Provision of System Components

You should determine the design of software components, and the compatibility between components from different suppliers.

F. Final Implementation Design

Reaching the final design involves trade-offs and compromises in many areas. Each aspect of the implementation may have its own priorities. The responsibility of the system implementer is to reach a final implementation design that meets the functional requirements while satisfying economic and organizational requirements.

VIII. OBTAINING LABORATORY PERSONNEL INPUT: NEEDS ASSESSMENT

A needs assessment should include operational review of the organization, departmental requirements, management needs, analysts' needs, and financial group needs, among others.

3

Data Management and Basic LIMS

Functional Requirements and Features

All Laboratory Information Management Systems (LIMS) must provide certain basic functionality: sample login, sample tracking, data entry, quality control/quality assurance, and reporting. Sample login involves entering information about the sample: who, what, where, when, and how. This means: Who submitted the sample, what analysis needs to be performed, where (in what department) is the analysis taking place, when should the sample be completed by, and how is the analysis going to be performed. After this information is entered, the LIMS must allow results to be entered for the sample together with appropriate quality control (QC)/quality assurance (QA) data. Finally, a report combining the sample information with the results is generated. Although most LIMS provide these basic functions, they typically perform them in different ways, some more efficiently and intuitively than others.

Many LIMS features are common to a wide range of laboratories. Requirements of a typical LIMS systems in an analytical testing laboratory might include the following:

- Sample login
- Sample tracking/barcode support/quoting
- Scheduling
- Chain of custody
- Instrument integration
- Result entry/audit trail
- QA/QC/specification checking
- Result reporting
- Web integration/links to enterprise software

- Chemical and reagent inventory
- Personnel training record tracking/instrument maintenance
- Archiving/data warehousing

This chapter will review basic functions that are fairly standard in most LIMS systems. Each feature and its purpose will be discussed. This section will also show the complexity of LIMS. Extensive thought and expertise on the part of the LIMS vendor are required to develop a system that can be used easily by various types of laboratories.

I. SAMPLE LOGIN FUNCTION

The single most important point of success of a LIMS is the sample login function for sample tracking. Sample Login must be quick and easy, capture all the required information, and yet be user-friendly. An onerous or complicated login can be a common point of failure of a LIMS, as this is often a bottleneck in many laboratories.

This function should capture all the relevant information for your laboratory, yet be set up to take advantage of pull-down lists, hot look-ups, and have as many fields "pre-filled-in" as possible. Limiting the number of keystrokes can save a great deal of time. Many laboratories incorporate bar-coding and scanners to pull information from the barcode on a sample label. There are several kinds of login fields, including text or numeric, or a combination.

There are two basic types of configuration for sample login: sample- or batch based. Sample-based systems assign information identifiers (such as client, collection information, etc.) to each sample. Batch-based LIMS assign information identifiers to a batch identifier. Each sample is then related to the batch identifier. This eliminates the need to enter duplicate information for multiple samples. Both types of systems allow testing information to be assigned to a sample. Sample numbering can typically also be user-defined or system-generated.

For certain types of fields, pick lists are ideal. For example, for sample delivery to the laboratory, the end-user may have a pull-down list with choices that include FedEx, USMAIL, UPS, courier, DHL, and others. Limiting items to list is very important when you want to query the database. For example, if you would like to know how many samples were received via FedEx, you can query by that field. If LIMS users were permitted to fill in the field themselves, you may find many entries in the database for FedEX: Federal Express,

Fed Ex, and so on. This causes the database to grow unnecessarily, puts "dirty data" into the LIMS and yields incorrect queries (any alternative spelling of FedEx would fail to identify relevant samples). By forcing users to select from a pull-down list you can ensure that queries will retrieve all the relevant records. You also have the capability to incorporate validity checking: for example, phone numbers for clients in The United States are limited to 10 number fields.

A popular field type is the automatic look-up field. You may type in a customer's or physician's name and all the associated contact information is pulled into the screen from the contents of another table. For example, when Dr. B. Cerius is entered in the LIMS, the name of the practice, address, phone, fax, e-mail, and billing information are automatically filled in. This is a huge time saver, especially when data entry technicians are logging in hundreds of samples or specimens each day. A good LIMS design will allow users to tab from field to field and allow use of the arrow function keys. The goal is to limit keystrokes and pull in data entry information from other sources where it may exist. Sample login is the portion of the LIMS that typically requires the most information to be entered; if this is cumbersome, it may be a bottleneck for the laboratory.

For larger pick lists, such as customers or samples sites that may consist of several hundred entries, a "hot look-up" might be more appropriate. In this case, the user need only begin typing the first few letters and the cursor will move to similar items in the database; the correct item can quickly be identified. Another feature of many LIMS is the ability to "limit to list." For example, if you select a sample type, based on that sample type the pull-down list will only allow the tests that can be performed on that sample. This helps to keep the lists manageable. In hierarchical pick lists, the contents of the second field are dependent on the contents of the first field. For example, if the user selects a sample type, the tests available for that sample type are displayed in the pull-down list of the second field. Only the database administrator should maintain the pick list, adding additional items as needed.

Another significant function built in to many LIMS systems in the ability to perform calculations. For example, when samples are logged into the LIMS, the "due date" for the analysis can be calculated automatically. This may be based on the date the sample is logged plus the holding time for a particular test. In another example, if you enter a birth date in a field, the corresponding age field may automatically be filled in with the correct age. Calculations can range from simple additions, multiplication, or divisions to complex queries.

II. TYPICAL INFORMATION COLLECTED AT LOGIN AND LOGIN FIELDS

The login of samples provides traceability for downstream analyses: who performed what tests, when, what the results were, if there were any special comments, and so on. The information collected can typically be organized into three categories: demographic information, operations information, and financial/billing information, which are defined here.

A. Demographic data

This include all the fields that provide information on where the sample or specimen was collected, whom it was collected from, the sample type, and how it was received in the lab. This may range from information shared among laboratories for each sample, to information unique to the sample, and collection information and terminology common to a particular laboratory. Most LIMS offer "user-definable fields" so that each laboratory can utilize their specific terminology, which translates into faster utilization and implementation.

B. Operational information

This includes data on which tests should be performed on the sample, whom the test results should be delivered to, what the actual results are, any warning flags, and when the analyses were completed. These fields encompass the custody and audit trail.

C. Financial/Billing Data

Since the features of an accounting system are complex, and LIMS are not accounting systems, the data in these fields are typically limited to simple invoicing and management reports. Information in these fields includes price lists, client-specific pricing, any special charges, and any other sample-related costs.

The information for these categories can vary significantly from laboratory to laboratory depending on the type of analyses performed. The type of information will also vary based on laboratory type, from research and development (R&D) to environmental to manufacturing to clinical laboratories. However, the overall goals are typically quite similar: to manage information better, decrease turnaround times, and improve the quality and reporting of data.

Some LIMS also collect information on the sample's condition upon arrival at the laboratory and allow users to add comments into the record, for example, that the sample was damaged during transit. Once the sample has been logged into the LIMS, a unique identifier given to it and tests have been assigned, the LIMS inserts a place holder for the test'(s) results in the database.

III. SAMPLE TRACKING, BAR CODING, AND QUOTING

Sample tracking allows users to locate a sample in the laboratory from login through analysis to final reporting and sample disposal. It includes the ability to list samples, identify their location, and identify what actions (sample preparation, analysis, interpretation, etc.) need to be completed. Samples can be retrieved by a variety of criteria including which are waiting for analysis, which are being analyzed and which have been completed. In addition, most LIMS include pre written, (canned) reports that can be printed automatically to provide information that the laboratory requires on a daily basis: production reports, backlog reports, worklists, turnaround time reports, and so on.

In addition to providing chain of custody information, sample tracking also provides the user with a status report on the sample: where the sample is physically located in the laboratory (which department), how long it has been in each department, and which analysts have handled the sample. Because laboratories are in the information delivery business, it is critical that users be able to determine the exact status of any sample at any time throughout the analysis-reporting process.

Most LIMS support bar codes (see sec. IV) within the sample tracking functionality. Places in the laboratory that bar codes can be utilized include sample collection, sample login, sample custody, storage and disposal, generating work lists, and automatic result entry with instrument integration. Each laboratory should determine its need to implement automatic identification, based on their specific sample management requirements.

Because a laboratory collects a wide variety of information on samples or specimens, the LIMS must be able to handle multiple data types. Data types include numeric values, text, comments (combinations of text and numeric), and image files. In addition, the LIMS must also be able to handle limits, date/time stamp, and user identification information.

Numeric values (whole numbers or fractions, typically represented by decimals) are probably the most common values saved in the LIMS. These are typically the result of analytical measurements, instrument files, and/or calculations.

IV. UTILIZATION OF BAR CODES AND AUTOMATIC IDENTIFICATION TECHNOLOGIES

More and more laboratories are utilizing bar code technology to assist in streamlining laboratory functions, thereby increasing productivity and accuracy. Bar code technology also allows laboratories to increase the amount of information available on the sample label, storing both text and numeric values. This information can easily be uploaded to the LIMS, thus avoiding double entry. Because bar codes can greatly increase the overall efficiency of sample and data management they are reviewed in detail here.

Efficiently managed laboratories require that many technologies, often complex, come together in a series of events. Laboratories that implement bar codes find that sample accessioning and tracking can be much smoother due to the more rapid and accurate use of bar codes. There are many different applications of bar codes in the laboratory environment. Laboratories may bar code sampling sites, sample containers, chemicals for inventory, reagents, material for disposal, and laboratory reports. Because not all laboratories operate the same way, a close look at the processes used in a particular laboratory will reveal where the use of automatic identification makes the most financial sense. This section describes the technology available and offers guidance in implementing automatic identification.

Automatic identification is used in a variety of business applications other than laboratories, including inventory control, work in progress, point of sales, credit card transactions, accounts receivable, marketing and business reply cards, biometric applications (voice and fingerprint recognition), and time and attendance. Many industries are replacing keyboard entry with bar codes. Bar coding is about 20 times faster and 20,000 times more accurate than keyboard entry.

Bar codes are relatively easy to implement. Any information can be converted to a bar code. For example a laboratory number 2000082438 can be visualized as a Times Roman or Arial font, if this is convert to a barcode font such as code 39 font the number will appear as a barcode. Bar codes (hardware and software) can be implemented for under a few hundred dollars. The typical return on investment occurs in under 1 year and the benefits to the organization make the investment in automation well worth it. Some benefits include increased productivity, faster speed, improved accuracy, and enhanced problem resolution.

Many laboratories agree that bar codes are a good idea, but are not sure where to begin implementing them. When considering implementation of automation software tools, the best advice is to phase in change and to address one area at a time. The following outline may be helpful: in deciding how best to implement bar codes in your laboratory.

First identify all of the processes of your laboratory business in a flow chart. Next, identify a specific laboratory procedure that would best lend itself to automatic identification. Then determine how and when to read (charge coupled device [CCD] or laser scanner) the bar codes. Finally, determine how (decide on font, printer type) and where to print the barcodes.

There are many different types of barcode systems; the use of any one is typically determined by the particular application. In its basic elements, however, a simple scanning system consists of a scanner, a decoder, a computer with barcode font, and a printer. Each element is described below.

A scanner is an instrument that reads the bar code. All scanners utilize a light-emitting diode (LED) and a photodetector to scan bar codes, creating a digitized signal that is sent to a decoder and converted to ASCII (computers readable), characters. There are many different input devices and the application will determine the scanner of choice. The types of scanners include hand-held wands that use CCD technology, laser scanners, and fixed focus optics (FFO). A decoder performs the conversion/translation of bars and spaces to ASCII characters. A software decoder performs the decoding utilizing the central processing unit (CPU) of the host personal computer (PC). However, decoding can also be performed within the scanner itself.

Other scanning devices that are becoming increasingly important in the modern laboratory include those using optical character recognition (OCR), intelligent character recognition (ICR), and biometrics identification (see sec. V).

OCR and ICR are often used in place of bar codes for automated data entry applications. The fonts are electronically scanned and digitized into ASCII characters. This is ideal for handwritten documents, where human readability is required and bar codes are impractical.

Computers have bar code fonts loaded into them, as any other font. There are several bar code fonts to choose from: the most commonly used in industrial bar code systems is Code 39. The primary feature is to encode messages using the complete alphanumeric character set. Two other popular bar code fonts include Code 128 and (universal product code-A (UPC-A). Code 128 is the most easily read with the highest message integrity; this results from its having several separate message check routines. UPC is a fixed-length code and the most commonly used for retail product labeling.

The printing techniques available include dot matrix, laser, thermal, and direct printing techniques. The best print quality is obtained with either thermal or direct techniques. Laboratories considering on-site printing must examine the costs of printing, start-up, and operation. These costs include personnel, computers to support the printing, printers, and consumable supplies (i.e. labels, ribbons, cartridges, etc.)

V. BIOMETRIC IDENTIFICATION

The Food and Drug Administration's (FDA's) ruling 21 CFR Part 11 has caused managers to take a closer look at security and also biometric identification (ID). Before we examine biometric ID, we need to understand traditional identification schemes. Access to privileged information or data is typically controlled with user-specific identities: user names, passwords, or personnel identification numbers (PINs). Some employers may use tokens, such as identification cards or so-called smart cards. The FDA rule recognizes the traditional security measures but requires such authentication schemes to be supplemented with work instructions and behavioral procedures that take reasonable care to protect the privacy and confidentially of passwords or tokens as not to compromise the system's security.

There are many inherent risks associated with traditional identification techniques such as passwords: passwords can be stolen by spying, easy passwords can be guessed, or they can be "hacked" via a dictionary-based security attack program. Passwords can be lost or stolen, or compromised because they are written down and left unsecured. In addition, people often reuse passwords for different systems, leading to further compromise. However, in the cases outlined above if the system does become compromised, new passwords can be issued, and new tokens can be distributed.

Biometric identification provides a higher-level authentication mechanism to ensure security of access and data. Biometric ID technology involves the ability to digitize an individual's physiological trait and use it as a means of identification. As the Part 11 legislation was developed, biometric ID was not common and lacked support infrastructure. However, the rule does allow for the integration of technologies as they become fully developed, supported, and available. Because most of the FDA-regulated industries are thought of as "closed systems," biometric ID is not required. An open system could be described as one in which a company is responsible for the content of data/

electronic records but it does not control access to those systems. Such a company could not comply with Part 11, without some form of digital signature and data encryption to ensure security and confidentially. See Figure 1 for a description of measures that should be taken to ensure data quality and integrity.

Biometric ID is used in high-security areas to protect restricted information or data. Biometric verification techniques include recognition of fingerprints retinal scans, voice patterns, and blood vessel pattern, in wrists or hands. The most frequently used is fingerprint identification. Table 1 lists a variety of available biometric identification options and describes the principle behind each. Before laboratories select a biometric ID method, several factors must be considered; accessibility trough laboratory protective equipment, cost, and environment (goggles, gloves, hazards inherent in chemical or biological laboratory).

There are also risks associated with biometric identification and they can be more severe than those associated with traditional security methods. Because we are all unique, we are the "password." This makes our "password" quite difficult to steal or imitate, but not impossible. Most attackers focus on "stealing" the digital imprint typically stored in the security input mechanism or associated computer software. The danger is that, unlike having a password stolen, you cannot simply get a "new" unique identification. Once the user's biometric data are stolen, they are compromised for the life of that user. A second concern is the technical implementation of biometric schemes. There is variability in biometric data, the equipment must be calibrated, and tolerances must be carefully monitored to ensure integrity of the system. A final concern is individual privacy; since individuals have only one unique ID, their work identity would not end at work.

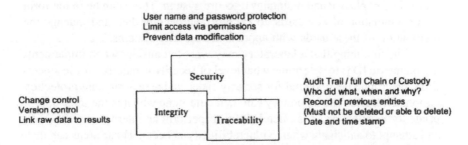

Fig. 1 Appropriate measures should be taken to ensure the security, integrity and tractability of electronic data, particularly in the regulatory arena.

Table 1   Primary Types of Biometric Identification Techniques Available

Biometric Target	Primary Feature
Finger scanning	Basically a fingerprint or thumbprint scan. This relies on the analysis of minutia points: finger image ridge (verification) endings, bifurcations, or branches made by ridges.
Finger geometry	A three-dimensional image of the finger is captured using a camera.
Hand geometry	Similar to finger geometry: a camera captures a three-dimensional image when a hand is placed on a hand reader.
Palm scan	Similar principles utilized in finger scanning by using minutiae found in the palm.
Signature analysis	Based on the characteristics that make each signature unique; movement of the pen, pressure applied, angle and other individual attributes.
Iris recognition	Analysis of the iris pattern; the colored ring tissue that surrounds the pupil of the eye.
Retina scan	Examination of the retina (layer of blood vessels located at the back of the eye) for pattern recognition. This is considered the most accurate of all biometric ID techniques.
Voice recognition	Unique characteristics of voice as a merger of physical and behavioral characteristics.
Face recognition	Analysis of unique shape, pattern, and positioning of facial features. There are two methods of capture: video cameras and thermal imaging.

Source: * Adapted from McDowall, 2000.

Several biometric alternatives can be utilized to ensure that the laboratory is in compliance and maintains a secure system. These can be in the form of a combination of key cards, to encrypt and decrypt data and manage the distribution of these cards with appropriate security systems.

The first thing that a laboratory may want to consider when implementing biometric ID is to determine what level of security is required for its operations, determine the potential for security risks, and assess the data protection options available. The laboratory can then determine which is the most appropriate for its situation. It is often helpful to perform an internal security audit and attempt to anticipate where vulnerabilities can occur. Those areas can then be fortified with cameras, additional passwords, and overall enhanced security. Laboratories can also turn to consulting firms or external auditors (who may come from other departments within the company) to access the "strength" of their security systems.

VI. ENTERING RESULTS DATA

Before generating a report, an LIMS must provide a means to enter results for an analysis. This can include both manual means and automated methods. Results are usually entered by analysis.

Following sample login, samples typically move through the laboratory from department to department for preparation, analysis, reporting, validation, approval, and final reporting. Once the sample or specimen is logged into the LIMS and given a unique identifier, and has passed through sample receiving and has had a test performed on it, results are entered into the LIMS. Although this is often entered manually by the analyst, it can also be entered electronically via direct instrument integration into the LIMS. Even when the data are automatically "down-loaded" into the LIMS, an analyst must still review the data to see if they are within specifications and to ensure that the proper controls were analyzed. Direct instrument integration has a rapid ROI and results in a reduction of transcription errors. They make sense especially when the instrument to be interfaced is prolific in its output (such as a GC/MS or ICP). With the popularity of spreadsheets and the fact that many LIMS users are familiar with a homegrown "spreadsheet LIMS," many result entry screens mimic this look, which is quite efficient.

The LIMS naturally monitors the movement of the sample throughout the laboratory and maintains a complete chain of custody. This is particularly important in forensic laboratories, since it identifies where the sample is physically located in the laboratory at any given time (e.g., refrigerator 1, shelf 2, with date and time stamped on it showing when it was checked in and out).

VII. REVIEWING AND APPROVING LIMS RESULTS

Data entered into the LIMS must be accurate and valid. Checking these data is often a multistep process. Users must first set acceptable limits for test results; these are typically categorized as "soft" warning limits and "hard" absolute limits. Once these result limits are set up for each test, client, project, or other variable, the LIMS will automatically warn the user if limits have been exceeded. Soft limits usually issue a warning that the user can still accept the result value; and hard limits issue a warning that the LIMS will not accept the data. Results are automatically checked against established specifications upon data entry and flagged if limits are exceeded. A person other than the analyst (usually a QC office or laboratory manager) reviews the results in conjunction with the associated QC test results to approve the data.

VIII. AUDIT TRAIL

Once a result has been entered into the LIMS, reviewed, validated, and approved, any changes to that result must be audited in accordance with Good Automated Laboratory Practice (GALP). Most LIMS will "lock out" users from changing the result after the validation and approval process. The only way to alter a result, following approval, is to spawn an audit trail. This process is straightforward. In most LIMS, users click on an audit button that asks them for certain information, such as the new result, and a reason for the change in the result. The LIMS will keep a rolling log of the old result (person who entered that result, date, and time), the new result, the reason for the change to the original result (with the time and date of the change, and the name of the person who made the change). The time and date stamp and the login name are typically autofilled in by the LIMS based on user login ID and the system clock. End users should not have permissions to modify the system clock or any other functionality on the server.

IX. RESULT REPORTING

Result reporting is the ultimate purpose of the LIMS: laboratories are in the business of information delivery in the form of final reports to customers. Whether the customer is another department in the same company or an external client, both need rapid, correct, and valid test results in order to make decisions, on their business or processes. Test results can be reported is a variety of ways: on a sample-by-sample basis, department by department, or by an entire project. The client usually dictates how they would like to see the data presented: tests listed vertically, results horizontally, graphically, or in a spreadsheet. Modern LIMS systems must have built-in query builders and the flexibility to accommodate the various permutations of required reports or export capabilities for electronic reporting.

4
Data Management and Advanced LIMS
Functional Requirements and Features

I. SAMPLE SCHEDULING

An extremely useful function is the ability to schedule samples in the Laboratory Information Management System (LIMS). Not all laboratories have the luxury of knowing which samples will arrive for what testing and when: but for those who do, automatic scheduling reduces workload and ensures that no samples are overlooked. Two types of samples can be entered into the LIMS: login and pre-login. Samples that arrive at the laboratory for analysis are logged into the LIMS, termed "login samples." Routine samples that the laboratory expects to receive can be prelogged in, termed "pre-login." These samples are then entered in the LIMS as pre-login or "pending," tests are assigned, bar code labels are generated for collection containers, and work lists printed. Once the laboratory receives the samples, the status is updated to login. This functionality is ideal for routine monitoring, such as that often encountered in wastewater testing.

II. CHAIN OF CUSTODY

The chain of custody tells an LIMS user or auditor exactly when the sample was logged into the LIMS system, who handled the sample throughout the analysis process, and where it was stored (e.g., lab bench A, refrigerator 1, shelf 2, box a, row 5) throughout the testing process. This is accomplished using time and date stamps and user IDs. An audit trail is required to generate a chain of custody in an LIMS. Everyone is responsible for the integrity and

35

custody of the sample as it passes through the different sections or departments of the laboratory. A chain of custody report details the specifics, and is extremely import in forensics laboratories.

III. INSTRUMENT INTEGRATION

The integration of instruments to LIMS offers significant benefits (speed, accuracy) and should be integral to any LIMS implementation. In the past, it was considerably more difficult to interface instruments to LIMS. Vendors typically could be classified into three categories:

1. Instrument vendors who wrote output files from their instruments to function only with their LIMS (thus providing a sales advantage)
2. Vendors who used proprietary software that made it difficult for anyone to interface their instrument to any LIMS without their involvement
3. Vendors who attempted to provide some standardization with a generic output

Most modern instruments provide common output formats such as CSV (comma separated values), Microsoft Access or Excel spreadsheets for their output format. This greatly simplifies the integration of instruments to LIMS. Some vendors will still encrypt the instrument output file and then sell clients additional software to unencrypt the output file so that the instrument can be interfaced. Prior to purchasing new instruments, customers should inquire as to the format of the output file, the ability to interface the instrument to a LIMS, and if any additional software must be purchased.

Electronic data transfer from analytical instrumentation to the LIMS replaces the keyboard entry of test and quality control (QC) result data. The data transfer involves the LIMS reading a file (often placed in a directory that the LIMS scans). The instrument file is typically "parsed" or massaged to be in a format that ensures that data are moved to the correct table in the LIMS. The data download is user-configurable. It can be configured in several ways: to import a single data file at a time or to import an entire data directory. Once the data are placed into the database, the LIMS can apply the same validity and limit-checking that the data would pass through if entered manually. An analyst can also choose to review and validate the data and QC results manually. The sample status of the imported results can be updated manually or automatically. If results are out of range, the LIMS should flag these samples and provide analysts the ability to delete the run before "bad" data is entered into the LIMS. When importing multiple determinations for a test, the LIMS should be flexible enough to allow analysts to delete individual determinations or the entire run. If data files are available in electronic format, a parsing file should be written to convert them into a format recognized by the LIMS to avoid transcription errors. If an instrument is on a local area network, the data can be transferred directly to the LIMS over the network. For instruments that are not networked, electronic data transfer is still the best method, albeit via sneakernet.

IV. QUALITY CONTROL AND QUALITY ASSURANCE

A. Quality Control

Quality control can be defined as the operational techniques and activities required to maintain and improve product or service quality. When most people hear the term QC in relation to LIMS, they think of the statistical quality control charts that have the values of sampling data over time, with upper and lower warning limits shown on the graph. These graphs provide excellent feedback to users on how the process or product is performing and establish confidence levels. Some users are interested in performing sophisticated statistical determinations on the data. Because most LIMS cannot handle these analyses, users often export their data to a specialized statistical analysis package such as SAS. It is important to select a LIMS based on open architecture and Open Database Connectivity (ODBC) compliance so that it can communicate with other applications. Users should avoid LIMS that use proprietary databases.

B. Quality Assurance

The LIMS typically contains a great deal of information on laboratory operation, data quality, and performance. In spite of this, few users, effectively mine the information in their LIMS. Many LIMS contain query builders'; screen on which users can check boxes for information they are interested in retrieving. For example, a laboratory manager can obtain information on each analyst; sample volume that they analyze per day, per month, per year, by test, by client, by the number of audits they signed off; and so on. Users can also examine workload by department, by instrument, or review QC data, and turnaround times for each department (from receipt in the laboratory through final reporting). By measuring overall laboratory performance, laboratory managers can identify areas for improvement and also commend areas that are performing well.

An LIMS can play a significant role in operations overall quality. Many of the reports generated from the LIMS, such as the analyses reports, statistical process control (SPC) charts, and trend analyses provide significant insight into overall product quality.

C. Specification Checking

Specification checking confirms that a material conforms to properties as defined by the consumer of the material. For example, for a raw material such as oil, the laboratory may measure the viscosity. There can be several different specifications for the same product based on the customer's requirements and uses of the product. LIMS are ideal for automatically notifying an analyst if a product does not meet certain specifications. They can also provide a match to customers whose specifications are met, so that the material may be used in another process by another customer. This is widely used in manufacturing industries.

There are several LIMS quality assurance programs, which include ISO 9000 international standards, government-regulated quality assurance programs, and guidelines such as the Environmental Protection Agency's (EPA's) Good Automated Laboratory Practice (GALP), and The American Society for Testing and Materials (ASTM's) guidelines. Other government agencies that participate in regulated quality assurance (QA) programs include the Food and Drug Administration (FDA), the Public Health Service, the armed services, General Services Administration, and state and municipal agencies. Of all the programs offered, the most comprehensive is probably the ISO 9000 certification, which requires biannual or annual audits, depending on the customer's request.

V. RESULT REPORTING

Following their validation and approval, results can be output to the client for final report generation. Reporting can be in formal or informal, in paper or electronic format, or as electronic data deliverables (EDD). Numerous types of reports can be generated by the LIMS; canned reports (these reports come with the LIMS), report generators to allow users to generate their own reports (Access, Crystal reports), spreadsheet reports, word processor reports, and graphical result reports. A few examples of reports that can be generated from the LIMS are listed below. Types of reported generated in any of these formates include chain of custody reports, workload or production reports, backlog reports, turn around time, instrument loading reports, and accounting reports (invoices).

VI. WEB INTEGRATION/LINKS TO ENTERPRISE SOFTWARE

Since the advent of the Internet, LIMS will never be the same. Now users demand and vendors deliver systems that can allow users and their clients to access their data remotely (from anywhere in the world) over secure networks, 24 hr a day and 7 days a week. Initially many LIMS vendors offered "LIMS Explorer-type" products that would allow access to the LIMS via the internet with a username and password. The next step Web-enabled LIMS. These are browser-based systems and many are rental systems; the LIMS vendor stores client data at their site, on their server.

Today more than ever, companies are focusing on total quality management (across their company) and are working toward integrating their operations. The tremendous success of SAP confirms this trend. LIMS must also communicate with other enterprise systems, such as accounting/financial packages, SAP, People Soft, SAS, and material requirements planning (MRP) systems, to name just a few. Most of the proprietary LIMS are being replaced with ODBC-compliant databases, such as Microsoft Access, SQL Server, Oracle, Informix, Sybase, and others as laboratories move to maintain their competitive edge.

VII. CHEMICAL AND REAGENT INVENTORY

Many LIMS have expanded their functionality to offer users more features that enhance laboratory productivity. One such feature is chemical, reagent, and even supply inventory. These systems are configured to allow users to set up their tests and record how much of each chemical or reagent is consumed for each test. The LIMS will automatically calculate the quantity on hand by performing reconciliation after each analysis. It will provide the user with a warning box when levels of certain chemicals and reagents are low. The warning limits and lead time for each chemical are user-definable. By keeping an electronic record of each vendor, ordering information (catalog number, quantity, grade, amount, cost, shelf-life, material safety data sheet [MSDS] information, any special shipping and handling instructions, etc.) and even a link to a specific vendor's Web site. This function can significantly expedite ordering and will save time in reordering. The major advantage of this functionality is the tight integration with the LIMS and automatic warnings.

VIII. PERSONNEL TRAINING RECORD TRACKING/ INSTRUMENT MAINTENANCE

Employee training records can also be stored in the LIMS. If an employee's training has expired, and if the LIMS has recorded current training as a required field, that analyst will not be allowed to enter information into the LIMS until the training is updated. The LIMS can also be configured to provide warnings as the training for specific users is about to expire (for example, at 4- and 2-month intervals). These types of reminders can be sent to the analyst and the laboratory manager simultaneously via e-mail, and help the laboratory to stay in compliance and maintain its employees' certifications.

Analytical instrument maintenance records can also be stored in the LIMS. Storing all instrument repair, replacement, or routine maintenance information in the LIMS can be invaluable for troubleshooting. The LIMS can also examine the performance of the instrument on control samples over time so that users are alerted to any potential problems with an instrument before it becomes a major problem. The LIMS can allow instrument maintenance to be a required field in the software, so that if the instrument is out of calibration users must either have the instrument calibrated or use another one before they can enter results for a particular test. This provides quality assurance that data entered into the LIMS were not generated by instruments out of calibration. This also provides the end-user with a comprehensive summary of instrument performance over time and early warnings to any problems that might arise.

IX. ARCHIVING AND DATA WAREHOUSING

Archiving refers to the process by which "old" data that have previously been reported and are no longer examined can be removed from the "active" database and moved to a different database. The archived data are, however, still readily accessible. There are several reasons for archiving.

1. To clean up a LIMS database: The laboratory may have retired par-
    ticular test methods and all associated analyses (previously re-
    ported) performed by this method. This ensures that the database
    only includes current methods and avoids confusion.
2. To enhance system performance: A decrease in system performance
    may result from large amounts of information (stored in tables) be-
    ing transferred over a network with limited bandwidth.
3. Time limits: The laboratory may wish to archive data based on cer-
    tain time periods, such as annually or based on government (regula-
    tory agency) timelines for data retention.

Whatever the reason, archiving should only be performed on samples
that have passed through each department and for which final reporting has
been completed. Most LIMS will not allow database administrators to archive
active data. Database administrators should maintain an archive log according
to the laboratory's quality assurance plan.

X. DEVELOPING A BACKUP PROCEDURE FOR YOUR LABORATORY'S LIMS DATA

Everyone knows that backing up data is important in the event that the system's hard drive crashes, a virus infects your system, or there is a catastrophic
event. But not many people are prepared to recover all of their data in this
situation. Because data backup is such an important issue, it is discussed here
in some detail. All laboratory database administrators should establish a solid
backup procedure to minimize the impact of lost or corrupted data. This section will focus on factors to consider when preparing a backup strategy for
your valuable laboratory data. Users must consider the time, effort, and resources extended to obtain that data in the first place, and plan their backup
strategy accordingly. In addition to planning in a networked environment,
the database administrator must consider the importance of backing up only
the data tables on the server or perhaps server configuration information. In
certain cases workstations acting as instrument controllers may also require
backup, especially if users are unable or for some other reason are not placing
critical files onto the network server.

There are many good commercially available backup utility software
programs. A backup utility functions to archive data from a hard disk to a
removable medium, of which there are many choices. Some of the most popular include small- and large-capacity removable disks, various types of mag-

netic tape, and optical disks. Small-capacity removable disks are very popular due to their low cost and ease of use. They typically hold less than 500 MB, are popular for workstation backups, and many newer computers have these *tape backups* built in. The large-capacity removable disks are similar to the small capacity disks and can hold more than 500 MB. They are unable to back up an entire server on one disk, although they are ideal for full workstation backups. Table 1 examines the evolution of storage devices.

Removable optical disks are very popular and come in several varieties: compact disk read-only media (CD-ROM), CD read (CD-R), read-write, (CD-RW) and digital video display (DVD). These optical disks utilize a laser or other type of light source to read and write information stored on it. The advantage of optical disks is their reasonable cost, longevity, and their capacity (650 MB to 17 GB). Optical jukeboxes contain hundreds of disks and can hold terabytes of information. There is also the so-called magneto-optical backup medium that combine the best of magnetic and optical storage. Tape backups have been in existence the longest, have the best-established infrastructure for support, and are often the most inexpensive. Some common formats include Travan, digital audio tape (DAT), digital linear tape (DLT), and quarter-inch cartridge (QIC). Popular back-up media are outlined in Table 2.

A. Scheduling Backups

There are many different ways to implement system backups. The three main types of backups are full, differential, and incremental. It is important to select the proper backup procedure and medium for the laboratory's system configuration.

In a full backup all information is backed up regardless of whether it has been backed up before, and the archive is cleared.

In an incremental backup, In only the "new" files or files that have been modified since the last backup are backed up, and the archive is cleared.

When a differential backup is performed, all files that were created or modified since the last full backup are backed up, and the archive is not cleared.

The fastest backup is the incremental backup combined with a regularly scheduled full backup. The disadvantage is that in the event of a crash, to restore the system you must restore the last full backup and every incremental backup (in the same order that they were backed up). Restoring this type of backup schedule can be time-consuming. In some cases a differential backup followed by a full backup may be more advantageous. Each database administrator must examine all the options and determine which backup option makes the most sense given his or her individual system hardware, software, and operations.

Some laboratories prefer to perform differential backups in combination with a regularly scheduled full backup. The advantage is that all new or modified files (since the last backup) are included; you would only require two sets of tapes to perform a full backup. Some laboratories opt for full backups. Even though these are the most comprehensive, they require the longest time to backup the system fully. On the plus side, the full backups are the fastest to restore. Backups should be performed when no files are open and preferably at night when no users are logged into the system.

Tape rotation strategies include, daily backup, weekly backup (also referred to as the Tower of Hanoi method), and the grandfather method. The daily method is best used when less than 25% of the data changes on a daily basis. In this method, three sets of tapes are used and on the first day of the work week a full backup is performed on the first tape set. For succeeding days in the work week, the second set of tapes is used to backup modified files, adding each day's changes to the end of the tape set. At the end of the work week, another full backup is performed, the tapes are stored off-site, and the second tape set is erased.

Another strategy is a weekly backup, in which tapes are labeled for each day of the week that the laboratory operates. The advantage of this method is that it is simple to follow; a disadvantage is that you cannot store more than one business week in the past. In the Tower of Hanoi method, five sets of backup media are used with the first being used every other day, the second being used every fourth day, and the third every eighth day, and the fourth and fifth, every sixteenth (alternately). This strategy can be used with daily or weekly backups. Finally, the Grandfather method uses one tape set for each of the first four days of the work week, one tape set for each of the end-of-week backups, and one tape set for the end of the month backup. This results in four media sets for each day, Monday through Friday, which are rotated each week. Four media sets are required for each Friday of the month and are rotated every month, and 12 media sets are required for the end of each month that are rotated each year.

B. Testing and Storing Backups

The best backup strategy, perfectly executed, is useless unless it is tested. Database administrators should develop a procedure and routinely restore data from the backup tapes. In addition many software utilities have programs that include data verification features to be used to ensure the copy's validity. Some

Table 1 Evolution of Storage Devices

Year	Size (Inches)	Disk Space	Description
1956	24	160 K	IBM introduced the 350 disk storage unit, the first random-access hard disk. The size of a dishwasher, it had 50 two-foot platters that held 5 million bytes of data (approx. 1,500 typewritten pages)
1973	8	17.5 MB	IBM released the 3340 (Winchester hard disk) with a capacity of 70 MB spanned over four platters.
1980	5.25	1.25 MB	Seagate introduces the first 5.25 inch hard disk.
1985	5	650 MB	First CD ROM drive introduced on PC's with read-only capacity.
1988	2.5	10 MB	Release of the Prairie Tek 220, the first 2.5 inch hard disk for portables
1998	5	2.6 GB	Debut of the DVD-RAM drive with rewritable capacity
1999	1	340 MB	IBM's new Micro Drive, for portable devices.
2001	3.5	10 GB	Quinta's optically assisted Winchester (OAW) drive, expected to store 20GB of data per square inch.
2005	3.5	58 MB	Typical desktop hard disk that holds 280 GB on five platters.

Source: Adapted from PC Magazine, 1999.

laboratories choose to perform the backup on a nonproduction computer to allow time to check the readability of the data. If an extra computer is available, this is an excellent idea.

Backup tapes should be stored in a safe location, preferably a locked, fire-proof safe. Another copy should be stored off-site. Each laboratory is different and its storage strategies will depend on how critical the data are and the cost involved in obtaining them. Commercial firms offer media storage for a fee; the obvious disadvantages are the cost and the time required to have the backup copy delivered.

C. Glossary of Terms: Tapes Used for Digital Back Up

Digital audio tape (DAT) was originally designed as the next generation of audio tape, providing compact disk (CD) quality sound in a cassette tape for-

Table 2 Popular Backup Media

Medium	Category	Capacity
Floppy diskettes	Small-capacity removable	1.44 and 2.88 MB
SyQuest cartridges	Small-capacity removable	33,88,200,230 MB
Iomega Zip disks	Small-capacity removable	100 and 250 MB
Imation SuperDisk	Small-capacity removable	120 MB
Removable hard disks	Small-capacity removable	Variable
SyQuest cartridges	Large-capacity removable	1 and 1.5 GB
Iomega Jaz disks	Large-capacity removable	1 and 2 GB
Removable hard disks	Large-capacity removable	Variable
CD-ROM, CD-R, and CD-RW	Removable optical	650 MB to 1 GB
MO disks	Removable optical	650 MB, 1.3 GB, and 4.6 GB
DVD	Removable optical	4.7 GB to 17 GB
QIC	Magnetic tape	100 MB to 25 GB
Travan	Magnetic tape	400 and 800 MB; 1.6, 2.5, 4, and 10 GB
DAT	Magnetic tape	2, 4, and 12 GB
DLT	Magnetic tape	35 to 70 GB
Standard 8 mm	Magnetic tape	3.5 to 14 GB
Mammoth	Magnetic tape	20 and 40 GB
AIT	Magnetic tape	25 and 50 GB

mat. DAT utilizes a 4 mm cartridge that conforms to the digital data storage (DDS) standard utilizing helical scan recording. This similar to the manner in which videotapes are recorded. Uncompressed DAT cassettes can hold 2–12 GB data.

Digital linear tape (DLT) is magnetic tape that utilizes one-half inch, single-hub cartridges. DLT is an adaptation of reel-to-reel, in which the tape cartridge performs as one reel and the tape drive as the other. DLT cartridges utilize the widest tape and data are recorded in a serpentine fashion, on parallel tracks grouped in pairs.

Quarter-inch cartridge (QIC) is one of the oldest tape formats, and is available as 3.5 inch minicartridges and 5.25 inch data cartridges. The QIC capacities range from 100 MB to 25 GB.

Travan evolved from the QIC format and provides higher capacity by utilizing wider tapes, different tape guides, and improved magnetic media. It is ideal for small, peer-to-peer workgroup backups. Its capacities vary, but typically range from 400 MB to 10 GB.

The 8 mm format is similar to the DAT tape, that both utilize helical scan recording but 8 mm tends to provide greater capacities. The 8mm tape, mammoth, and advanced intelligent tape can provide capacities of 3.5 to 50 GB.

The greater storage capacity now available has allowed software developers more space for their code. That is not always a good thing, however; in the past programmers were very conscientious about writing succinct code, whereas extra capacity means that queries and code did not have to be recycled. Programmes could relax and "fill" this excess space. This has been evident in the explosion of functionality and new features in operating systems alone. In 1981, Microsoft's disk operating system (DOS) required less than 160K disk space. Eleven years later, Microsoft introduced Windows 3.1, which was familiar to Macintosh users, and required 10 MB for a standard installation. Microsoft Windows 2000, released in 1999, required 100 MB for a full installation. As hard disks have become smaller and smaller, the amount of information that they contain has exploded.

5
Life Cycle of LIMS Software Development

Whether a laboratory information management system (LIMS) is purchased from a commercial vendor or a laboratory hires a consultant to write a database application, some design considerations are fundamental. This chapter will focus on the various design concepts and fundamental considerations necessary in constructing or purchasing an LIMS. There are several different software development approaches and they are all valid. However, some are more appropriate for simple projects while others may be better suited for complex systems. The development team can determine which approach is best suited for its particular project.

Database design concepts include the traditional or waterfall method, the Pharmaceutical Manufacturers Association (PMA) software development validation life cycle, and the object-oriented software development life cycle.

I. TRADITIONAL WATERFALL CONCEPT

The traditional method of software development consists of a cascade of tasks that begins with analysis, followed by design, coding, testing, and integration. In this early model there is a unidirectional migration from one task to the next, with no opportunity to go back and modify the design or code. This model is rigid and may not be the best choice for a dynamic environment such as the analytical laboratory.

47

II. PHARMACEUTICAL MANUFACTURERS
ASSOCIATION LIFE CYCLE

Because the pharmaceutical industry is highly regulated, they have defined
the LIMS life cycle in terms of the validation process, which includes system
hardware as well as software. The phases of the Pharmaceutical Manufacturers
Associations (PMA's) system life cycle are more rigid than the traditional
waterfall concept, primarily to satisfy the strict regulatory requirements of the
Food and Drug Administration (FDA). There is little flexibility between the
various phases of the PMA life cycle:

> Defining system requirements
> Design phase
> Programming phase
> Test phase
> Installation phase
> Operational phase

A. Defining System Requirements

In the first phase, the LIMS team assembles and defines the functional requirements of the LIMS. This phase is usually represented by input from several
different individuals and/or groups and is characterized by flow charts of the
entire processes. In the PMA model the functional requirements must all be
defined at the earliest possible stage. However, when dealing with complicated
systems, this is not always possible. Also, future refinement of initial design
typically results in a much better system.

B. Design Phase

In this development phase software engineers translate the system requirements and flow charts into screen designs (layouts) and begin to design table
structures. The team typically begins with high-level design and then breaks
the project into smaller pieces or modules to allow for detailed design. Once
the design plan is complete, it may be presented to end-users for additional
feedback prior to beginning the software coding to ensure that the software
still matches the system requirements. This phase typically requires sign-off
by the LIMS project manager. A solid design is crucial, since this will serve
as a blueprint for the entire system. The design is referred to as the entity

diagram because is shows all the relationships within the database. These entities comprise the table scheme.

C. Programming Phase

This phase is characterized by translation of the design into computer language or code and unit testing and is analogous to building a house based on accepted blueprints. A unit can be defined as a module, function, or subroutine. Unit testing is initiated so that specific pieces of the code can be tested before all the pieces are put together. This accelerates the software development and allows the software engineers to check, and perhaps optimize, the code before it in utilized in the core software. This development approach also allows anomalies to be determined and corrected in the early phases of development.

D. Test Phase

The test phase also marks the beginning of software validation. Although testing should be performed during each phase of development, the test phase focuses solely on testing. There are several excellent books on software testing and validation and there are several different approaches and strategies, depending on the type of system designed. It is typically best to test during the early stages of software development and then throughout the product development life cycle utilizing checklists, inspections, and reviews. Testing strategies include unit testing, module testing, integration testing, and on-line testing. Units make up modules, so unit testing is the first level of testing that can be performed, followed by testing larger sections of code or modules, which are comprised of multiple units. Integration testing examines the effects of the various modules operating together. Software tools are available to assist in the testing process. The final testing stage is on-line testing; continuous on-line testing allows users to evaluate the software thoroughly and uncover any anomalies.

E. Installation Phase

The main focus of the installation phase is validation of the system and the installation of the hardware. Hardware here includes the server, client machines, network, hubs, network cards, any other equipment or instruments interfaced with the LIMS for automated data migration. Close attention must be paid to vendor-recommended specifications for the client and the server, network protocol, and hardware design specifications, and to the manufacturer's recommendations for environmental conditions. The latter includes adequate ventilation in the server room, reliable surge protection, and an uninterruptible power supply for emergency operation in the event of a power failure.

F.   Operational Phase

In this phase the system "goes live." It is also the last stage in the software development life cycle. In addition to routine operation, maintenance is also performed and service packs are applied to correct any software anomalies. The PMA approach is quite structured. As a result, it may take longer to reach the operational phase than when the object-oriented Model, for example, is used.

III.   OBJECT-ORIENTED APPROACH

This method of software development consists of several phases including needs analysis, design phase, software evolution, and continued modifications. It is one of the most flexible development approaches and perhaps the most successful for complex systems such as a LIMS, since it involves early prototyping and continual opportunities for enhancements and modifications. Although a controlled development cycle is preferred from a regulatory standpoint, if the final product is obsolete or misses the mark, it will be of little value to end-users. Perhaps the most most useful aspect of this approach is the interaction of developers and end-users early in the development process. This is critical since different users will come to the project with their own terminology. They may think that they are communicating effectively when in fact developers do not fully understand their functional requirements. Because this method involves producing rapid prototypes, end-users are made aware of discrepancies early on in the project and avoid serious delays later. The four elements of the object model are described briefly here.

A.   Analysis

Analysis should begin with a meeting between the system developers and end-users, to determine the functional requirements and system specifications. Common analysis methods used include, entity–relationship analysis, structured analysis, and object-oriented analysis.

B. Design

The design phase is characterized by incorporation of the functional require-ments into the project considerations, including cost, development tools avail-able, use of existing or new hardware, internal expertise, network and op-erating systems, and any additional limitations.

C. Evolution

This phase is characterized by coding, testing, and integration of the software product. Users are permitted to provide continual feedback and the product is allowed to evolve slowly to what users desire. Developers can continually improve their software based on lessons learned early in the project or past experience. This is a popular development model since it mimics human cog-nition and is easy to modify.

D. Modification

This phase is marked by software upgrades, service packs, and enhancements. Although similar to the evolution phase, it is different in that marks the chal-lenges encountered in the laboratory environment and how the software can adapt to them. As users begin to become familiar with the software, they understand its power and request additional features and functionality. If the core LIMS is well written, it will have the flexibility to adapt to these new requirements.

IV. DESIGN CONSIDERATIONS

A. Graphical User Interfaces

With the introduction of the graphical user interface (GUI) by researchers at Xerox Park, users have been able to learn new systems quickly, because many common elements are shared by different programs. Microsoft Word, Excel, and Access all utilize features that can be found in many LIMS: scroll bars; radio buttons; bold, underline, and italic buttons; on-line help; pull-down menus; hot look-ups; open/close and minimize buttons; tile and cascade fea-tures, among many others. Because different programs share these common features, users can often learn or at least navigate within complex programs based on their previous knowledge. An extremely important element in select-ing a user-friendly LIMS is the interface, since many different users will be accessing the system. Simple, clean, and uncluttered screens are preferred over screens with excessive buttons or features that do not follow typical Windows conventions. The interface should include on-line help, search ability, and a tutorial for first-time users. A common misconception is that GUI systems require the use of the mouse; in fact, most can utilize either the mouse or the tab keys. Some LIMS also allow the assignment of functions to keys, for those reluctant to give up the "function key" concept from the days of character-based software and operating systems.

With dramatic advances in hardware and software, the additional graphic requirement of the GUI interface over the character-based systems is no longer an issue. Although the character-based systems are faster than GUI systems, the benefits of GUI (ease of use, faster implementation) far outweigh the speed. Even speed issues are lessening as advances in hardware continue. Just as there are very few remaining rotary phones today because the technology has moved to digital, there are fewer character-based LIMS each year. An intuitive interface is key to successful LIMS implementation.

B. System Help

The ability to place the entire LIMS manual on-line is a great help to first-time users. Key features of all LIMS should include context-sensitive on-line help that includes screen shots and definitions of new terminology, hot look-up, search abilities, a tutorial, hypertext links, and a glossary of terms to help new users master the new LIMS's functionality. Complete hard copy documentation of the LIMS should also be provided.

C. Source Code

Source code is not frequently supplied for several reasons. One reason is that it contains the intellectual property of the vendor that developed the program. Perhaps more important is that once the source code is modified, the end-user has a system unique to him or her. This makes support challenging and shifts the impetus for updating software (applying source code fixes) to the end-user. Although it may sound like a good idea, in reality the inherent support can become burdensome. Many people purchase commercially available LIMS so that they do not have to provide end-user support, issue service packs or product enhancements, or keep the software current by adding features and functions. With a commercially available system, end-users can share ideas, reports, and code with other users in the user-group setting. Technology can move further faster than is possible if the laboratory is working with its own limited resources.

D. Migration from a Manual LIMS

End-users will face many challenges as they migrate from a manual or paper/ spreadsheet system to an automated LIMS. However, if the LIMS is well designed to accommodate the laboratory's sample flow and data management requirements, the greatest challenges will be simply learning the new system. The LIMS must also be flexible enough to accommodate new requirements of data storage and reporting (regulatory or business) faced by the laboratory. End-users often find it difficult to "trust" the LIMS with data and calculations previously performed manually. However, once users understand and experience the benefits of a relational database, they often request additional functionality. In learning any new system, there must be a commitment from management to adopt it, encourage additional training, and allow users to invest the time to learn and understand it.

6
Regulatory Requirements

Laboratory accreditation demonstrates a laboratory's competency to perform certain processes and tasks consistently. Unfortunately there are many different accreditation programs. Each type of lab—research, testing, or manufacturing—must comply with certain quality assurance regulations. A Laboratory Information Management System (LIMS) must meet the requirements of the laboratory's quality assurance programs. Some of the better-known regulations and standards for ensuring quality are International Standards Organization (ISO) Standards, US Environmental Protection Agency (EPA) Standards, and US Food and Drug Administration (FDA) Standards.

These programs assess and certify the capability of various types of laboratories to conduct testing. However, the requirements and assessments vary considerably according to the organization and program. Some programs are quite comprehensive; others only involve minimal review of a laboratory's qualifications. In addition, many of the accreditation programs are industry or application-specific.

Laboratories can be accredited in several ways: to test in an entire field of testing, or in a scientific discipline such biochemistry, or in a specific technology such as gene splicing, or in relation to specific products such as pharmaceuticals. Because most US laboratory accreditation programs are designed to meet particular governmental or private-sector needs, such programs tend to take distinctive forms and use different sets of procedures and criteria to ensure that a laboratory has sufficient competence to perform the specified testing. Some programs require only a simple review of data submitted by a laboratory, while others require an on-site evaluation of the laboratory's facilities, staff, equipment, quality system reviews, and proficiency testing.

I.   ISO 9000

The International organization for standardization (ISO) standards must be complied with by products sold in Europe. ISO 9000 was conceived as an international standard and was quickly adopted by the European Union as a means of standardizing quality processes across national boundaries in Europe. More than 100 countries have endorsed the standard, although they generally give it a different alphanumeric designation. In the United States, it is known as American National Standards Institute (ANSI)/American Society for Quality Control (ASQC) Q90.

The ISO 9000 series of standards defines what a quality system should do, but not how to do it. That is left up to the individual company. Each company must meet certain requirements. Section 4 of the standard contains the quality system requirements detailing the 20 areas of required conformance, which are summarized here.

- Management responsibility: Management must define, implement, communicate, and maintain quality objectives and assign personnel at all levels of the organization to be responsible for verifying the company's quality system. Periodic management reviews are required.
- Quality system: Creation and implementation of a quality manual, including documented procedures and instructions.
- Contract review: The company must document customer orders and verify that it can meet customer requirements.
- Design control: Documentation of quality measures in design, including design planning, input, output, verification, and changes.
- Document control: Procedures for creating, distributing, and tracking all documents, including changes related to ISO 9000 activities to ensure the use of only the latest revisions.
- Purchasing: Establishes procedures for supplier assessment, selection, review, and monitoring as well as verification of purchased product quality.
- Purchaser-supplied product: Stipulates how a company should handle, store, and maintain customer-supplied materials.
- Product identification and traceability: Specifies how the company should identify products through all stages of production, delivery, and installation.
- Process control: Formalizes and ensures controlled conditions for

production and installation processes that affect quality. Requires document control and maintenance.

- Inspection and testing: Establishes procedures for inspection and testing of incoming, in process and outgoing products.
- Inspection, measuring, and test equipment: Defines the requirements for equipment maintenance and calibration.
- Inspection and test status: Establishes a system for identifying the status of products as they move through the facility.
- Control of nonconforming product: Ensures that products not conforming to requirements are prevented from inadvertent use or installation.
- Corrective action: Establishes, documents, and maintains procedures for investigating nonconforming products and initiating preventive action.
- Handling, storage, packaging, and delivery: Formalizes procedures for product handling, storage, packaging, and delivery.
- Quality records: Establishes procedures for identifying, collecting, indexing, filing, storing, maintaining, and disposition of quality records.
- Internal quality audits: Requires regular internal quality audits to ensure compliance with ISO standards. Requires corrective action of deficiencies.
- Training: Identifies and provides training for all personnel performing activities that affect quality. Requires training records for verification.
- Servicing: Where servicing is specified in a contract, requires procedures for performing and verifying service activities.
- Statistical techniques: Specifies the use of appropriate statistical techniques for verifying process capability.

ISO 9000 regulations primarily affect manufacturing laboratories. Laboratories for companies that are ISO 9000 certified will have established quality assurance practices that comply with the ISO 9000 regulations. It is important to consider how the laboratory's quality assurance (QA) protocol will be implemented in the LIMS. For ISO 9000 laboratories, items to consider include, but are not limited to, the following:

- Management responsibility: What security level within the LIMS will be assigned to personnel, logging in samples, entering data, validating data, approving data and editing data?

- Quality system: How can standard operating procedures be handled in the LIMS?
- Document control: How will the LIMS handle revisions to test methods?
- Product identification and traceability: What information about the product will be required to be logged into the LIMS?
- Inspection, measurement, and test equipment: How can the LIMS record information on the calibration of equipment?
- Quality Records: How will the LIMS obtain and process quality records? How will the LIMS produced certificate of analyses, certified lab results, charts, and graphs?
- Statistical Techniques: What statistical techniques will the LIMS utilize to produce control charts, process capability charts, and trending charts?

II. ISO 25

ISO 25 differs from ISO 9000 in establishing a laboratory's technical competence, ISO 9000 focuses on conformance to a quality system. ISO 25 requires criteria specific to the goal of ensuring valid test data. It states that

> in addition to periodic internal audits, the laboratory shall ensure the quality of results by monitoring test and/or calibration methods. . . . using statistical techniques, proficiency testing programs, certified reference materials, replicate tests using this aim or different methods, read testing, correlation of results for different characteristics of an item. The selected methods should be appropriate for the nature of the work undertaken.

ISO 25 is the most widely recognized lab accreditation standard. Many of the National Environmental Laboratory Accreditation Conference (NELAC; see Sec. VI) standards are based on ISO 25.

An LIMS can accommodate the accreditation requirements. Section 2.11.2.2 states that laboratories must have procedures for recording and reporting relevant data and operations relating to sampling. These records must include, the sampling procedure used, the identification of the sampler, environmental conditions and diagrams or other equivalent means to identify the sampling location as necessary, and, if appropriate, the statistics that the sampling procedures are based upon.

The technical requirements section of the ISO 25 standard describes criteria that the laboratory has to meet in order to demonstrate that it is technically competent for the type of tests and/or calibrations it undertakes. Section 3.3.4.3 states that "the suitability of the methods may be taxed and confirmed by comparing the method with specified requirements typical for the intended use. The range and accuracy of the values obtainable from this method as assessed for the intended to use shall be relevant to the customer's need." A LIMS can address this requirement by automatically flagging out-of-range results.

III. GOOD AUTOMATED LABORATORY PRACTICES

Most of the health and environmental data that EPA uses in its regulatory programs are analyzed in and reported by laboratories. Laboratories are increasingly utilizing LIMS to aquire, record, manipulate, store, and archive data.

Good Automated Laboratory Practices (GALP) are a union of federal regulations, policies, and guidance documents establishing a uniform set of procedures to ensure the reliability and credibility of laboratory data. GALPs are EPA's response to mounting evidence of corruption, loss, and inappropriate modification of computerized laboratory data by EPA contractors. They are applicable to all EPA organizations, personnel, or agents of the EPA who collect, analyze, process, or maintain laboratory data for the agency. Laboratories and organizations utilizing an LIMS that wish to improve the integrity of data are encouraged to implement and comply with applicable GALP provisions.

GALPs supplement the good laboratory practices with government and EPA policies that address automated hardware, software development and operation, electronic transfer, and system security, specifically: laboratory management, personnel, quality assurance unit, LIMS raw data, LIMS software, security, hardware, comprehensive testing, record retention, facilities, and standard operating procedures. The GALPs' objective is to provide EPA with assurance of the integrity of LIMS raw data. According to section 8.4.2, individuals responsible for entering and recording LIMS raw data are to be uniquely identified when the data are recorded, and the time and date are documented. The LIMS software must be able to record this data whether data are entered manually or electronically transferred from laboratory instruments. If raw data are transferred from laboratory instruments into the LIMS, the date and time of data transmission and the name of the specific laboratory instrument must be recorded. The LIMS must be able to document a change to the raw data. The system must provide an audit trail that provides clear evidence that a change was made, explains the reason for the change, records the date of the change, the person who made the change, and the person who authorized the change.

GALPs have a regulatory impact that varies considerably based upon the ultimate reviewer of the data generated. Pharmaceutical and biological laboratories are not subject to GALP requirements, although FDA investigators use GALPs as one of their guideline documents. The contract laboratory program uses GALPs as a criterion in contract renewal. Whether compliance is required are not, GALPs provide a functional and operational definition of the kinds of controls necessary to provide confidence in the data generated by a laboratory.

GALPs are based on six principles:

1. The system must provide a method of ensuring the integrity of all entered data. Communication, transfer, manipulation, and storage/recall all have the potential to corrupt data. Demonstration of control necessitates the collection of evidence to prove that the system provides reasonable protection against data corruption.
2. The system's formulas and decision algorithms must be accurate and appropriate. System users cannot assume that the test or decision criteria are correct. The formulas must be inspected and verified.
3. An audit trail that tracks data entry and modifications back to the individual responsible is a critical element in the control process. This generally utilizes a password system or its equivalent to identify the person or persons entering data, and generates a protected file to log all unusual events.
4. Also crucial is a consistent and appropriate change-control procedure capable of tracking the system operation and application software. All software changes should follow carefully planned procedures, including a preinstall test protocol and appropriate documentation update.
5. Control of the most carefully designed and implemented system will be thwarted if appropriate user procedures are not followed. This principle implies the development of clear directions and standard operating procedures (SOPs), the training of system users, and the availability of appropriate user support documentation.
6. Consistent control of the system requires the development of alternative plans in case of system failure, disaster, and unauthorized

access. The principle of control must extend to planning for reasonable unusual events and system stresses.

IV. ELECTRONIC SIGNATURES

On August. 20, 1997, the 21 Code of Federal Regulations (CFR), part 11, went into effect. This is more commonly known as the FDA final rule on electronic records, signatures, and submissions. The final rule provides criteria under which the FDA will consider electronic signatures to be equivalent to traditional handwritten signatures. Electronic signature that meet the requirements of the rule will be considered to be equivalent to full handwritten signatures, initials, and other general signings. The FDA is encouraging industry to move away from paper records and submissions towards electronic records and forms. However, it is crucial to be sure that the electronic signature on an electronic record can be traced to a single authorized individual. The electronic signature must be linked to the respective records so that the signature cannot be excised, copied, or otherwise transferred to falsify an electronic record by ordinary means. Electronic signatures must be unique to one individual and must not be reused or reassigned to anyone else.

Laboratories that implement an LIMS will find that many systems have the foundation for validating electronic records and signatures in accordance with part 11, but not all conform to the final rule in a precise manner. 21 CFR 11 applies to records in electronic form that are "created, modified, maintained, archived, retrieved, or transmitted" under any record requirements set forth in regulations by the FDA, including electronic records submitted to the agency under requirements of the Federal Food, Drug, and Cosmetic Act and in the Public Health Act.

A LIMS is considered a closed system in which access is controlled by persons who are responsible for the contents of electronic records in it. 21 CFR 11 requires the employment of procedures and controls designed to ensure the authenticity, integrity, and (when appropriate) the confidentiality of electronic records, and that the signer cannot readily repudiate the assigned record as not genuine. Procedures and controls shall include but are not limited to the following:

- Validation of the LIMS to ensure accuracy, reliability, consistent intended performance, and the ability to discern invalid or altered records.

- Limit access to authorized individuals.
- The ability to generate accurate and complete copies of records in both "human-readable" and electronic form suitable for inspection, review, and copying by the FDA.
- Use of secure, computer-generated, time-stamped audit trails to record independently the date and time of operator entries and actions that create, modify, or delete electronic records. Record changes shall not obscure previously recorded information.
- Use of authority checks to ensure that only authorized individuals can use the system input or output device, alter records, or "perform the operation at hand."

Electronic signatures may be based on biometric devices or at least two distinct identification complements, such as an identification code and password. A biometric device verifies an individual's identity based on measurement of his or her physical features or repeatable action, given that those features and/or actions are both unique to that individual and measurable. Biometric devices include iris scanners, fingerprint scanners, palm print scanners, and retinal pattern scanners. The devices are connected, usually by direct wire or network, to the system that requires authentication and are typically placed next to the computer. Biometric devices verify identification by comparing patterns presented by the individual at the time of verification with stored patterns obtained from the authorized user under controlled circumstances. Electronic signatures based on biometric devices must be designed to ensure that they cannot be used by anyone other than the genuine owner. Some devices integrate biometric signature capture with cryptographic technology to bind signatures to documents with date/time stamps. Other biometric-based electronic signature systems use dynamic signature verification with a parameter code recorded on a magnetic strip card.

Laboratories that use electronic signatures based upon use of identification codes in combination with passwords must employ controls to ensure their security and integrity. The controls must have the following capabilities, but are not limited to the following:

- Maintain the uniqueness of each combined identification code and password, such that no two individuals have the same combination of identification code and password.
- Ensure that identification code and password issuances are periodically checked, recalled, or revised (e.g., to cover such events as password aging).

- Use of transaction safeguards to prevent unauthorized use of the passwords and/or identification codes and to detect any attempts at their unauthorized use.

An LIMS can conform to the FDA rule on electronic signatures through the use of passwords and biometric devices, which include palm print readers, fingerprint readers, and iris scanners. The FDA does not establish numerical standards for levels of security or validation. Although the FDA requires operational checks, authority checks and periodic testing of identifying devices, it is up to the laboratory to determine suitable methods to accomplish these tasks. Currently only the FDA regulates electronic signatures but it can be expected that other federal agencies, such as the EPA, will soon follow suit.

V. GOOD MANUFACTURING PRACTICES

Manufacturing and pharmaceutical laboratories must comply with current good manufacturing practices (cGMP). The GMPs are a group of requirements found in the legislation, regulations, and administrative provisions for methods to be used in, and the facilities or controls to be used for, the manufacturing, processing, packing, and or holding of a drug. The GMPs were developed to ensure that a drug meets the safety requirements, has the identity and strength it purports to and meets the quality and purity characteristics represented as possessing. GMPs are the part of quality assurance ensuring that products are consistently produced and controlled to quality standards. 21 CFR parts 211–226 contain the cGMP requirements as regulated by the FDA for drugs and parts 600–680 contain those for a biological product for human use. GMPs cover the following provisions:

- Organization and personnel
- Buildings and facilities
- Equipment
- Control of components and drug product containers and closures
- Production and process controls
- Packaging and labeling controls
- Holding and distribution
- Laboratory controls
- Records and reports
- Returned and salvaged drug products

Aspects of the regulations pertinent for LIMS implementation are found in 21 CFR 211.194 (Records and Reports). The laboratory records must include complete data derived from all tests necessary to ensure compliance with established specifications and standards including examinations and assays. A LIMS utilized by manufacturing or pharmaceutical laboratories in compliance with GMP should comply with the 11 provisions for laboratory records outlined below:

1. A description of the sample received for testing with identification of source, quantity, lot number or other distinctive code, date sample was taken, and date sample was received for testing.

2. A statement of each method used in the testing of the sample. The statement must indicate the location of data establishing that the methods used in the testing of the sample meet proper standards of accuracy reliability as applied to the product tested. The suitability of all testing methods used must be verified under actual conditions of use.

3. A statement of the weight or measure of sample used for each test.

4. A complete record of all data security in the course of each test, including all graphs, charts, and spectra from laboratory instrumentation, properly identified to show the specific component, drug product container, closure, in-process material, or drug product, and lot tested.

5. A record of all calculations performed in connection with the test including unit of measure, conversion factors, and equivalency factors.

6. A statement of the results of tests and how the results compare with established standards of identity, strain, quality, and purity for the component, drug product container, closure, in-process material, or drug product tested.

7. The initials or signature of the person who performed each test and the date(s) that the tests were performed.

8. The initials or signature of a second person showing that the original records have been reviewed for accuracy, completeness, and compliance with established standards.

9. Complete records must be maintained of any modification of the established method employed in testing. Such records must include the reason for the modification and data to verify that the modifi-

cation produced results at least as accurate and reliable with material being tested as the established method.

10. Complete records must be maintained of any testing and standardization of laboratory reference standards, reagents, and standard solutions.

11. Complete records must be maintained of the periodic calibration of laboratory instruments, apparatus, gauges, and recording devices.

VI. NATIONAL ENVIRONMENTAL LABORATORY ACCREDITATION CONFERENCE

The National Environmental Laboratory Accreditation Conference (NELAC) is sponsored by the United States Environmental Protection Agency (EPA) as a voluntary association of state and federal officials. The purpose of NELAC is to "foster the generation of environmental laboratory data of known and documented quality in a cost-effective manner through the development of nationally accepted standards for environmental laboratory accreditation." The NELAC standards are applicable to the following EPA statutes: the Clean Air Act (CAA), the Comprehensive Environmental Response Compensation and Liability Act (CERCLA), the Federal Insecticide, Fungicide and Rodenticide Act (FIFRA), the Clean Water Act (CWA), the Resource Conservation and Recovery Act (RCRA), the Safe Drinking Water Act (SDWA), and the Toxic Substances Control Act (TSCA). The NELAC standards incorporate ISO 25, ISO 43, and ISO 58 standards.

An LIMS can be a significant tool for a laboratory to comply with the NELAC standards. The requirements of the quality systems are clearly defined in Chapter 5 of the NELAC standards.

A. Equipment

An LIMS can maintain records for major equipment including the name of the equipment, manufacturer, serial number, date received and placed in service, current location, dates and results of calibrations, maintenance details, and repair history.

All measuring operations and testing equipment having an effect on the accuracy or validity of tasks must be calibrated and/or verified before being put into service on a continuing basis. This includes balances, thermometers, incubators, volumetric dispensing devices, and controls standards.

B.   Calibration

The LIMS should be able to store the method-detection limit and the reporting limit, since results cannot be reported that are less than the reporting limit. It should also be able to track the initial and continuing calibration verification checks as a part of the quality control function.

C.   Computers and Electronic Data-Related Requirements

Section 5.10.6 of the NELAC standards pertains to computers or automated equipment used for the capture, processing, manipulation, recording, reporting, storage, or retrieval of tests data. The standard references section 8.1–8.11 of GALP. The LIMS must comply with the standards. There must also be a procedure for the maintenance of security of data, including the prevention of unauthorized access to, and the unauthorized amendment of, computer records.

D.   Sample Handling and Receipt

The laboratory must have a documented system for uniquely identifying the items to be tested including identification of all samples, subsamples, extracts and digestates. Each sample container received in the laboratory must be assigned a unique identification code, and the condition of the sample upon receipt must be recorded. The laboratory ID code is the link that associates the sample with related lab activities such as sample preparation or calibration. A LIMS must be able to record all of the required chain of custody information: unique laboratory ID code, sample location, date and time of collection, collector's name, preservation type, sample type, received date and time, receiver's name, and any special remarks concerning the sample. The requested analyses including approved test method numbers must be linked to the unique laboratory ID code assigned to the sample. Samples rejected by the laboratory must logged in to the LIMS and assigned a unique laboratory ID code. Comments regarding the rejection of the sample should be linked to the laboratory ID code.

E.   Record-Keeping System

NELAC requires that the record-keeping system (LIMS) must allow historical reconstruction of all laboratory activities that produced the resultant analytical data. The record must include the identity of all personnel involved in sampling, sample preparation, calibration, testing, and data verification. Electronic data must not be erased or overwriten. An audit trail can record any and all changes to the data and the person who made the changes. The original and the changed data must be maintained. Electronic data shall be easily retrieved and access to the archived data must be documented.

F. Laboratory Sample Tracking

The LIMS should be able to track the sample in the lab. Items that must be included are sample preservation, sample container, and compliance holding times; sample identification, receipt, acceptance or rejection by unique laboratory ID code; sample storage and tracking; sample preparation including procedures, ID codes, volumes, weights, and calculations; date of analysis, analysis type, and initials or signature of analyst; instrument identification; data and statistical calculations; quality control data, and electronic data security.

G. Laboratory Report Formats

NELAC requires that each report to an outside client must include the following: A title such as "Test Report", "Laboratory Results," or similar; the name and address of the laboratory, including a contact name and phone number; unique identification of the report, such as a serial number and the total number of pages; name and address of the client and the project name; identification of the tested sample; date of receipt of the sample, date and time of sample collection, dates and times of analysis; identification of the test methods used; any deviations from test method or conditions that may have affected the quality of results; measurement, examination, and arrived results; signature and title or equivalent electronic identification of person accepting responsibility for the report and the date of issue; results of tests performed by subcontractors must be clearly identified by subcontractor's name or accreditation number; and a certification statement that test results meet all requirements of NELAC or provide reason and or justification if they do not.

The regulatory environment is changing much more swiftly than it has in the past and requires much more activity on the part of manufacturers and laboratories. The intensifying regulatory requirements evolving from the recently revised GMP, electronic signature rule, and NELAC regulations are beginning to affect decisions regarding LIMS. The easier it is to validate a LIMS, the less work is required by the lab to maintain it. Since time is becoming an increasingly scarce commodity in the laboratory and elsewhere, a laboratory must ensure that any future LIMS purchases comply with all required regulations at installation.

7
Hardware and Operating System Requirements

A laboratory information management system (LIMS) is composed of more than the LIMS vendor's software program. It also includes computer hardware such as servers, mainframes and personal computers; peripherals such as data loggers, bar code readers, printers, tape drives, and scanners; operating system software including Unix, Windows 95, Windows 98, Windows NT, and Windows 2000; databases such as Microsoft SQL Server, Oracle, and Microsoft Access; reporting software such as Microsoft Access and Crystal Reports; laboratory instruments and networking components such as network interface cards, operating system, and network cabling hubs and routers. Each component plays a role in the success or failure of the LIMS.

I. COMPUTER HARDWARE

The original personal computer (PC) was introduced in 1981. In less than 20 years it has changed our means of communicating. All PCs share the same basic technology and consist of a central unit and various peripherals (Fig. 1).

A. Central Processing Unit

The central processing unit (CPU) is the "brains" of the PC, in some way responsible for every thing the PC does. It determines, at least in part, which operating systems can be used, which software packages the PC can run, how much energy the PC uses, and how stable the system will be. The processor is also major determinant of overall system cost: the newer and more powerful the processor, the more expensive the machine.

69

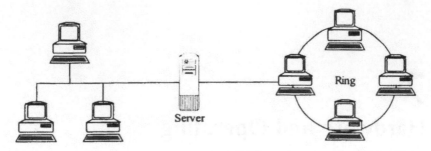

Fig. 1 Client/server computing.

PCs are designed around different CPU generations. Intel is not the only company manufacturing CPUs, but it is the leading one. CPUs have for years doubled their performance every 18–24 months; there are no indications that this trend will stop.

The CPU is centrally located on the motherboard. It continually receives instructions to be executed. The work itself consists mostly of calculations and data transport. All of the elements of the processor stay in step by use of a clock that dictates how fast it operates. Current CPUs have a clock speed of 450 MHz, which means that the clock "ticks" 450 million times per second. The higher the clock speed, the faster the CPU processes data. Many network servers possess two CPUs. When installing a new LIMS system, confer with the vendor to determine the CPU requirements of the software your system will be using.

B. System Memory

System memory is the place where the computer holds current programs and data that are in use. Demands made by increasingly powerful software have accelerated system memory requirements at an alarming pace over the last few years. System memory has three levels: primary cache, secondary cache, and random access memory (RAM). The primary and secondary cache are used for temporary storage of instructions and data organized in blocks of 32 bytes. Primary cache is the fastest form of storage, followed by secondary cache.

Random access memory (RAM) is the main memory area accessed by the hard drive and acts as a staging post between the hard drive and the processor. The more data it is possible to have available in the RAM, the faster the PC will run. RAM is built of rectangular arrays of memory cells with support logic in the arrays that is used for reading and writing data. There are several types of memory chips: Dynamic RAM (DRAM), Extended Data Out DRAM (EDO DRAM), and Synchronous DRAM (SDRAM). DRAM is the most basic and cheapest memory chip. EDO DRAM reads a theoretical 27% faster than the DRAM. Synchronous DRAM (SDRAM) is the newest memory chip architecture and its chip supplies the bits of data as fast as the CPU can take them. SDRAM is 18% faster than EDO DRAM. Memory has become less expensive recently. A simple rule of thumb is that the more memory a computer has, the better it will perform.

C. Hard Drive

When the power to the PC is switched off, the contents of its random access memory are lost. The PC's hard drive serves as a nonvolatile bulk storage medium and is where software applications are stored. As applications continue to increase in size, they require more and more hard drive space. Over the past several years the average size of a hard drive has risen from 100 megabytes (MB) to over 12 gigabytes (GB). As drive capacity has expanded, however, prices have gone down. Performance has also increased. The performance of a hard drive is important to the overall speed of a computer system. A slow hard drive has the potential to hinder a fast processor. Today's fastest hard drives are capable of an average latency of less than 3 msec, an average seek time of less than 7 msec, and maximum transfer rates approaching 20 Mb/sec.

Hard drives come in two basic standards: Enhanced Integrated Drive Electronics (EIDE) and ultra Small Computer System Interface (SCSI). SCSI has become the accepted standard for server-based mass storage. For servers, the drives are usually arranged in a Redundant ARray of Independent Disks (RAID) to provide both high speed and high availability. If one hard drive should fail, data is not lost because it is distributed among the remaining drives. The benefits of RAID outweigh the drawback that the storage capacity of one drive is lost. If four 4 GB drives are configured for RAID, you have a maximum storage capacity of 12 GB.

D. Tape Drives

To protect data, hard drives should be backed up on a regular basis. Backing up a hard drive is the process by which data contained on it is copied to a removable storage device. If a hard drive crashes, the backed up data can be restored to the system. Recordable CD ROM drives, JAZ drives, and tape drives are all considered removable storage devices. Tape has been the traditional security backup medium and remains the best choice based on capacity and cost. The more inconvenient a security backup regimen is to implement, the less likely it is to be used. With the size of the average hard drive now measured in gigabytes, tape is generally the only medium that allows a complete hard drive to be backed up without needing to swap media during the process. The most popular solutions for low-end systems are digital audiotape (DAT) and quarter inch cartridge (QIC) linear tape. Various media options and back up strategies were previously described in Chapter 4 in greater detail.

E. Optical Drives

Optical drives of various formats (optical reporting methods, CD recording formats, DVD recording formats, and magneto-optical recording formats) are beginning to gain acceptance: their cost is dropping relative to tape technology and their capacities and transfer speeds have increased dramatically in the past few years. Optical recording methods can be classified into two categories: recordable and rewritable. Recordable technologies produce read-only disks. Rewritable technologies allow the disk to be overwritten many times.

The original, most popular, and least expensive rewritable optical recording method is magneto-optical (M-O). M-O drives range in size from a few hundred megabytes to several gigabytes.

CD-R recorders use one of the write-once, read-many recording methods to create read-only CDs that can be read in any normal CD-ROM player. They are generally used to create permanent archives on bootable CDs. CD-RW recorders produce CDs that can be overwritten. Unfortunately CD-RW CDs are not readable in a normal CD-ROM drive. Most CD-RW drives can produce CD-R CDs that can be read in normal CD-ROM drives. Both CD-R and CD-RW drives have a very small capacity (680 MB) and a relatively slow transfer rate.

Digital video disk (DVD) is a new format full of promise. The different recordable formats can save 2.6–4.6 GB per side for up to 9.2 GB of storage per disk. The transfer rate is still slow compared to modern tape drives. There are several DVD formats: one read-only format and three rewritable formats. DVD-RAM disks are read only and are not readable in normal DVD drives. DVD-R are recordable drives that produce disks that can be read in any DVD drive, but they are expensive (around $15,000 US dollars) and in short supply. DVD − RW and DVD + RW are two different DVD rewritable formats that are currently incompatible and are supported by different vendors. Once the differences in recordable DVDs are settled, DVD might replace M-O as the most cost-effective archive solution.

F. RS-232 Port

The RS-232 interface is also called the serial port. It is the serial protocol in which a single transmitter sends one bit of information over a single communication line to a single receiver. The values for RS-232 communication parameters may vary widely among computers and instruments. The devices to be interfaced must use the same settings on both ends. The communication parameters are:

- Baud rate: transmission rate in bits/sec
- Start bit: signals the beginning of a transmission sequence
- Data bits: the actual data message in ASCII format
- Parity bit: optional bit that checks correctness of the data sent
- Stop bit: signals the end of transmission sequence
- Hardware handshake: receiving device is ready for transmission
- Software handshake: special characters used synchronize communication
- Communication pathway: unidirectional (simplex), bidirectional at different times (half duplex), or bidirectional simultaneously (full duplex)

Various software programs can initialize the communication parameters between the computer and any analytical instruments, and parse the data. The parsed data can be saved into a file that can be imported into the LIMS.

LIMS have been known to fail not because of software problems but because of hardware issues. The criteria for selection of a LIMS should be driven by the software function, but hardware considerations should not be ignored. Computer hardware technology and price performance ratios used to support LIMS are changing rapidly. A lab should start with vendor guidelines for sizing the computer hardware. These are often the minimum requirements necessary for the software to operate. To obtain optimal LIMS performance, a lab should be prepared either to upgrade their current computer hardware significantly or to purchase new, state-of-the-art computers during LIMS implementation. However, the latter may be outdated in 2–3 years.

Hardware sizing is dependent on many factors:

- Number of concurrent users
- Number of data records
- Archive requirements
- External loads on the system from non-LIMS applications

Elements to consider are CPU, clock speeds, system memory, hard drive capacity, archived media capacity, and network bandwidth. LIMS transactions often put demanding loads on the network and computer hardware.

II. COMPUTER NETWORKS

LIMS utilize a company's computer network to share and process information. Computer networks can be classified into two broad categories: local area networks (LAN) and wide area networks (WAN). A LAN enables data exchange between devices such as computers and printers within a small geographical area. A WAN is usually composed of a collection of LANs and spreads over large geographical areas. Networks are commonly discussed in the framework of the Open Systems Interconnect (OSI) model, which is a uniform standard introduced in 1978 by the ISO to facilitate data communication among the disparate equipment from different vendors. Figure 1 shows the conceptual layout of the OSI model. Each layer performs a specific set of network functions and builds on its immediate predecessor, resulting in progressively higher-level functionality through the seven layers.

A. Physical Layer

The physical layer is the lowest layer in the OSI model and deals with how electrical or optical signals are generated, transmitted, and received in a physical medium. The most common standard at this level is the RS-232C. The physical layer consists of four major components:

Transmission format; digital or analog
Physical transmission medium; electrical or optical signals
Data encoding; what signal patterns represent 1s and 0s and the synchronization between sending and receiving devices
Physical medium attachment; wiring and pin layout connectors

B. Data Link Layer

Networks transfer data in packets of a certain size. The data link layer forms packets, manages their movement at each node in the network, and adds the appropriate addresses of the source and destination nodes. In PC-based LANs, at the data link layer protocols such as Token Ring and Ethernet are implemented. These functions are often performed by network interface cards (NICs) installed on computers.

C. Network Layer

The network layer performs data routing; the physical path that data packets travel. Routing is based on network conditions, priority of data packets, and other factors. In a LAN each node communicates with all other nodes within that LAN. The data packets need not go through intermediate nodes to reach their final destination; the network layer therefore does not have a great deal of importance. It is important, however, in WANs, where intermediate routing is often needed. The network layer software usually resides on special switching nodes that are part of the network. Hubs and routers are part of the network layer.

D. Transport Layer

The transport layer makes sure that messages are delivered in the order in which they were sent. It also guarantees that there is no loss or error in the delivery of the messages. The size and complexity of a transport protocol depend on the type of service it can get from the network layer or data link layer. The common transport protocols include transmission control protocol/Internet protocol (TCP/IP) and sequence packet exchange/Internet packet exchange (SPX/IX).

E. Session Layer

The session layer is responsible for establishing virtual connections between processes running on different computers. Each connection is called a session. The session layer is usually the interface to the network for programmers. It handles such functions as name recognition, administration, and security in the network. A common protocol for the session layer is NetBIOS extended user interface (NETBEUI).

F. Presentation Layer

The presentation layer is responsible for translating one data format into another. At the sending computer, the presentation layer transforms the format sent by the application layer into a common format. At the receiving computer, the presentation layer translates the common format into a format chosen by the application layer. Terminal emulation occurs at this layer. At the presentation layer the network operating system (NOS) software such as Windows NT or Novell resides. The NOS software offers network services and provides the application layer with independence from lower-level details. Many applications, however, often bypass the NOS to interact directly with the session or transport layer, allowing for greater control over network resources.

G. Application Layer

The application layer serves as the user interface by which network resources can be accessed. The application layer offers the following functions:

- Resource sharing, such as sharing of printers
- Remote file access
- Remote printer access
- Interprocess communication support
- Remote procedure call support
- Network management
- Directory services
- Electronic messaging, including email

With interprocess communication (IPC), two different processes, such as applications, on different computers have the ability to communicate. In a PC-based LAN, IPC is provided by the network operating system. IPC is important to client/server computing because the client application must be able to communicate with the server application. IPC is implemented at various OSI layers including the presentation, session, and transport layers. Different levels of implementation affect system transparency and flexibility. High-level implementation of IPC offers high system transparency but less flexibility; lower-level implementation offers less transparency but more control and flexibility.

III. COMPUTING MODELS

A network is a very structured system with distinct modules running at different levels and performing various tasks that ultimately allow computers to communicate with each other. They are numerous computing models for ways in which computers can work together. Today's networks are often heterogeneous; composed of many client computers, mainframes, and file servers.

Computers electronically process and manipulate binary streams of data and have traditionally been grouped into three categories: mainframes, mini-computers, and microcomputers. Mainframes include Cray supercomputers, IBM 3090s and DEC Vax machines. The DEC Vax was once the mainstay for LIMS; it has essentially vanished being replaced by powerful minicomputers and microcomputers. Workstations, including those produced by Sun Microsystems and Silicon Graphics and the DEC 64-bit Alpha, belong to the minicomputer category. Personal computers (PCs) are usually considered microcomputers. The boundaries between the categories are becoming increasingly indistinct due to the rapid advances in semiconductor technology and microprocessor design.

A computer system composed of one or more computers is classified based on the system's collective processing power and how it is used by the end-user. Computer systems can be classified into four types: single processing, shared processing, file server computing, and client/server processing computing. Single processing, shared processing, and file server computing have been used for a long time. Client/server computing is a relatively new computing model but one that has become very important for LIMS.

A. Single Processing

Single processing is a computer used by one user. The computer has exclusive access to its own system's resources including the CPU, memory, and data. All the needs of the user (processing, storage and retrieval of data, user interface, and printing) are taken care of by one PC. The chief advantage of single processing computing is that it is simple. This scale of the single processing system is usually quite small and relatively inexpensive, and customization of the computer environment is much easier than in other computing environments. Single processing computing is impractical, if not impossible, when resources are to be shared by group of people. Information is shared using floppy drives and the so-called "sneaker net."

B. Shared Processing

Shared processing is designed to allow multiple users to use one machine, usually a mainframe, simultaneously. Multiple terminals are connected to a mainframe computer as shown in Figure 2. A user logs into the mainframe from a terminal and opens up a "session" with the mainframe. The terminal provides an interface through which a user enters commands, runs programs, and sends and receives data. Each session runs independently so that different

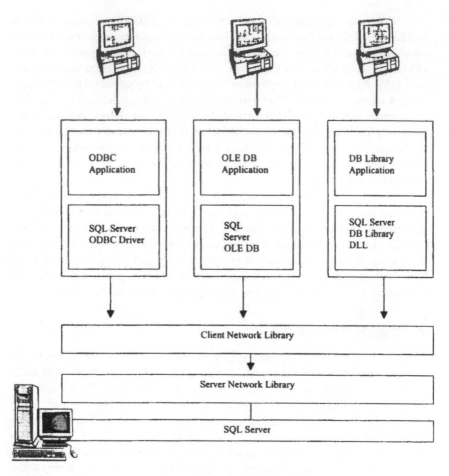

Fig. 2 Communications with client/server configurations.

users logged onto the mainframe do not interfere with each other's work. The mainframe keeps track of all the current users and sessions.

Mainframe terminals were originally simple interfacing devices specifically designed to communicate with the mainframe. They usually consisted of little more than a monitor and a keyboard. They did not possess any processing power and were called "dumb" terminals. Today the role of terminals is often taken by PCs running terminal emulation software. The processing power of the PC is not used.

Shared processing is a highly centralized computing method by which the mainframe stores and processes all the data. Since all the data reside in one place, it is equally accessible to all users; any changes in the data are almost instantaneously available to all users. The mainframe is easy to manage and allows for the easy imposition of security on the data. Mainframes are powerful and complex, often possessing several CPUs and large amounts of memory, but they are expensive. The application development cycle for mainframes is much longer than for other systems and can be expensive to maintain. Many early commercial and in-house LIMS were developed for mainframes such as the Vax, but few modern LIMS use mainframe-based shared processing today.

C. File Server Computing

Server architecture is a form of computing in which both data and computing resources are shared between two or more computers. File server computing uses file servers and clients and is sometimes referred to as client-based computing. All the data processing is done on workstations (clients). File servers hold shared data and answer requests from clients to perform certain simple tasks, including transferring files between workstations and file servers and issuing instructions to peripheral devices such as printers. When the central file server is located on the LAN, computers on the same LAN can share data as well as expensive peripheral devices such as laser printers. File server computing is simpler and less costly to implement and maintain than mainframe-based computing.

There are several important limitations to the file server architecture. When a workstation that is on a LAN requests a file from the server, the file server sends a copy of the entire file to the workstation using the LAN's network software. This is an especially relevant issue with database systems. For example, a user may only need to make some changes to one record in the database, but because the file server cannot selectively send data to a workstation, the user must have all the records in the database downloaded to the workstation. Only then can the user start to make changes to a record. The downloading of the entire database file is not only unnecessary but also can congest the LAN traffic and decrease overall performance of the database system. In addition, file servers "lock" a file so that no other workstation can access it until the first workstation is finished. Different users cannot access the same database file at the same time, although users may wish to access different records within it. File server architecture is not conducive to the smooth functioning of a LIMS, when large numbers of users require simultaneous access to the system.

D. Client/ServerComputing

Modern LIMS and enterprise computing hinge on a concept of sharing: information, resources, and processing power. As more and more computers are brought into a network to work together, they are organized into logical architecture so that each computer's role and function are clearly delineated. The client/server architecture, how firmly established, divides information-processing between front-end (client) components and back-end (server) components. By distributing the workload between front-end and back-end modules, client/server computing gives a system a great deal of flexibility as well as high performance.

Client/server computing arose in response to limitations in the single processing, mainframe-based processing and file server processing architectures. Client/server computing is also a form of distributed computing and is comprised of three components: client, server, and intercommunications between client and server components. The client/server architecture is well suited for database systems. Many database software products conform to the relational data model that is a framework for organizing data and provide efficient and powerful tools for structuring and manipulating data.

IV. OPERATING SYSTEMS

An operating system (OS) is a collection of system programs that together control the operation of a computer system. The operating system includes programs that:

- Initialize the hardware of the computer system
- Provide basic routines for device control
- Maintain system integrity in handling errors
- Provide for the management, scheduling, and interaction of tasks

There many types of operating systems, the complexity of which varies depending upon what type of functions are provided and what the system is being used for. A general-purpose operating system can run a number of different programs: games, word processing, business applications, and program development tools. The operating system's first job is to initialize the computer system's hardware before running application programs. Currently used operating systems include DOS, Windows 3.1, Windows 95, Windows NT, Windows 2000, and Unix.

Microsoft's disk OS (MS-DOS) was designed for the IBM series and compatible range of computers and was introduced in 1981 when IBM produced its first computer. Based on the 8088 microprocessor, the memory range was limited to 1 MB, 640K of which could be RAM while the other 640K was reserved for hardware devices such as video cards and expander cards. Computers rapidly advanced, but not until DOS version 5.0 was released was support above 640K of RAM available. MS-DOS is, however, still written in the old 8086 code format and does not take advantage of processors faster than 80386.

Windows 3.1 incorporates the graphical user interface (GUI). It increases performance over DOS alone by rewriting the file-handling routines as 16-bit code. The layout allows users to share files, clipboard contents, CD-ROM drives, and printers with other users.

Windows 95 is an upgrade to Windows 3.1 and is designed for use as a workstation client or desktop system. It is not intended to be used on a server, but can be used in simple workgroups to share resources such as printers and files. It is more reliable than Windows 3.1 and supports all major networking protocols. Windows 95 is capable of running 16 and 32-bit applications. It allows for plug-and-play and support for hardware devices, while still supporting existing MS-DOS and Windows drivers and programs.

Windows NT is designed for serious power users and desktop workstations. It has built-in data protection, supports common networks and protocols allows for preemptive multitasking, is scalable, and has support for more than one processor type. The software comes in workstation and server versions. It is designed for robust scalable networks based on domains. When servers are required to handle more than 10 clients, the NT server may be one of the best choices of operating systems.

A survey of LIMS users conducted in August, 1999, by *Analytical Consumer* showed that 87% of the laboratories that use an LIMS had a client/server architecture while the remaining 13% were using mainframe computers with PCs and terminals at their client locations. The power of the server varied with the size of the laboratory. Windows NT edges out Unix in its frequency of use as server software by a ratio of 10:4. Windows NT and Windows 95 are the main operating system choices for LIMS clients.

For many companies the LIMS extends beyond the laboratory and interfaces with megacontrolling systems such as SAP. LIMS is becoming part of an enterprise-wide information system.

8
Obtaining Laboratory Personnel Input

Laboratory Information Management Systems (LIMS) are complex systems that integrate hardware, software, people, and procedures. It is easy to focus on the hardware and software aspects of an LIMS to the extent that the people and procedural aspects are often overlooked. The LIMS is how a laboratory tracks and manages its information resources, particularly the data that represent the laboratory's product. Any change in the data-handling system, therefore, engenders some potentially traumatic changes in the way the laboratory operates. The laboratory staff is called upon to adopt new routines. The LIMS has to be compatible and integrated with the quality and business objectives of the laboratory.

The LIMS implementation team should be composed of representatives from each department that will be affected by the LIMS. Users, information services personnel, financial personnel, customer service representatives, clients, analysts, and managers all need to be involved from the beginning of the project. They have the best perspective on how the new system will work. Resistance to change may be the biggest obstacle to overcome when implementing a LIMS. Early involvement can help build user acceptance of the change.

The planning stage provides an opportunity to anticipate problems and educate the users on the project. Successful LIMS implementation requires a good understanding of the laboratory's operations and business practices. Planning involves clearly defining the information flow, structure of the generated data, and user requirements. Flow charts showing where and how data are generated, transferred, and stored in the system are very useful. These charts should include decision points in the process such as quality control data evaluations, supervisory approval, report generation, and invoicing. Figure 1 is an example of a workflow diagram for a laboratory. The workflow diagram

83

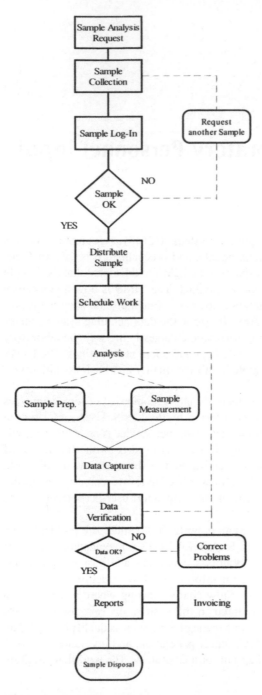

Fig. 1 Obtaining laboratory personnel input.

will elucidate the LIMS functions and interaction points with typical laboratory workflow (processing of samples). Specific laboratory requirements will vary from one laboratory to another. The following descriptions explain the basic LIMS functions and workflow interactions. The LIMS project manager will need to work closely with all laboratory personnel to develop a thorough and accurate workflow model.

I. LABORATORY WORK FLOW

A. Sample Analysis Request

The workflow process is initiated by a request for sample analysis. Examples of sample requests include manual forms, phone requests, time- or calendar-based requests, process-driven requests, and LIMS-generated requests. Information obtained from a sample request includes client information, sample information, required tests, and safety information.

B. Sample Collection

Sample collection may be a manual or automated process. The condition of the sample that comes into the laboratory should be able to be documented. Examples include notations of preservation, temperature of sample, and the condition of sample containers. Sample collection can be assisted by printing collection lists and generating labels for sample containers.

C. Sample Login

The LIMS assigns a unique laboratory ID number to each sample logged into the LIMS. The system can capture who submitted the sample, costs, test required, and priority of the sample.

D. Distribute Samples

The sample distribution process includes LIMS functions of work lists, sample routing, custody, and labeling. It is sometimes necessary to divide the sample for simultaneous analysis at different workstations.

E. Schedule Work

The LIMS automatically schedules work (tests) for each sample. Lab management can adjust sample priorities and reassign work as required. The LIMS

can add laboratory standards, control samples, and quality control samples to
the scheduled workflow. LIMS statuses are updated for each sample.

F. Analysis

Analysis involves multiple steps: sample preparation, sample measurement,
quality control samples, and data capture.

G. Sample Preparation

Most samples need some preparation before undergoing analysis. In some
cases preparation requires entering experimental data such as tear weight and
final weight from a balance.

H. Sample Measurement

Test results are the main output of the measurement process. Results may also
be obtained for blanks, standards, and instrument self-checks. The results of
sample analysis must be entered into the LIMS. Data may be entered manually
or through an electronic interface where data are transferred from the instrument directly into the LIMS. When results are entered into the LIMS, the
status of the sample is updated. Audit trails record information about each
LIMS transaction.

I. Verification and Correction

A laboratory may require that a qualified person review results. The LIMS
can show summaries of work done for review. Unusual or out-of-range results
can be flagged for greater scrutiny. Corrections to data can be made in the
verification step. Changes to results should be audit trailed. Results can be
approved, thus changing the sample status. Results can also be marked as not
acceptable and enter a retest loop or a resample loop.

J. Reports

Once test results are verified and approved, they can be reported to the customer. Reports can take a variety of forms: printed output, electronic mail,
and response to online queries. Different reports can be generated to meet
different requirements.

K. Interpretation

The laboratory exists to generate information for their clients. The LIMS can organize and configure results to make reporting and interpretation easier. Statistical routines can be used to determine trends and data can be shared across departments and business units for enhancing decision making.

L. Disposal of Samples

The LIMS can be used to track final sample storage, disposal, and waste removal, or it can alert end-users if any samples are to be returned following analysis.

II. MANAGEMENT FUNCTIONS

By collecting statistics and time stamps at various points in the process, the LIMS's management functions can prepare reports for laboratory managers such as the number of samples processed and turnaround times. This can help to identify peak demand, roadblocks, and other problems. Turnaround times can also be documented. Billing outputs are mediated in those laboratories that charge customers for work done. Instrument calibration and maintenance records can be maintained and reported by the LIMS. System management functions include backup and recovery, user maintenance, and archives. Permanent archives are prepared after all the work is done on a sample. The ability to read archives after LIMS software updates is an important consideration.

The LIMS may need to interface with software from other departments, such as management resource planning systems (MRPs) and enterprise resource planning systems (ERPS).

These are both company-wide software systems used in manufacturing plants as important tools for resource allocation and for determining material requirements. The systems track materials, keep inventory, generate bills of material, compile work orders, and manage scheduling. In the past, many companies have experienced problems integrating different applications. MRPs and ERPs are usually supplied by a single vendor such as SAP to eliminate integration problems. In some parts of the business, ERPs and MRPs programs often lack the functionality to cover the laboratory. The LIMS provides the ability to manage laboratory data.

In general, an MRP or ERP system will manage all product recipes, process flows, specifications, and sampling protocols. The LIMS will manage the laboratory data, sample flow, chain of custody, analytical methods, and good laboratory practice (GLP) processes. The main areas of integration between the MRP/ERP system and the LIMS are the latter's ability to log in samples and product lots in the laboratory automatically based on schedules and information in the MRP/ERP systems, and to transfer the appropriate results back to the MRP/ERP system for dispositioning.

Integrating LIMS and MRP/ERP systems can be daunting. The two systems must have a robust bidirectional communication link that ensures data integrity. The interface may or may not be a custom solution that requires modification with each implementation.

III. QUESTIONNAIRE

After identifying all possible LIMS users and their job activities, a questionnaire should be provided to the users. It should ask for specific information and comments on what features the user would like the LIMS to have and may include items on data entry, instrument interfacing, testing requirements, hardware requirements, user interface considerations, and quality control requirements.

IV. PAPERWORK

In addition to evaluating the workflow of the laboratory, the paperwork used throughout the lab should be evaluated. Types of paperwork include data sheets, log books, laboratory standard operating procedures (SOPs), reports, and miscellaneous others.

A. Data Sheets

A data sheet (also called a bench sheet) is a form that is filled in as data are collected while a test is being run in the laboratory. Data sheets are kept as part of the official raw data records and usually include areas for sample identification, raw results, calculated results, blanks, standards, and comments on the test or samples. Data sheets will aid in the design of the test method entry screen used for the input of test data into the LIMS.

B. Log Books

Log books may contain sample login information, test method, calculations, test results, instrument calibrations, and sample status. Examination of the log books may reveal other sample information that may need to be entered into a LIMS.

C. Reports

Existing reports may include certificates of analysis, work schedules, customer test reports, daily sample analysis reports, quality control reports, and backlog reports, and lab production reports.

Evaluation of current reports will give additional insight into the type of information that will need to be entered into LIMS, stored in, and retrieved from the LIMS. The examination of a sample analysis report may indicate that customer information needs to be stored in the LIMS. Test types and sample types will therefore need to be stored. Test results, both numerical and descriptive, must be able to be entered into the LIMS. All information must be able to be retrieved quickly.

V. LABORATORY INSTRUMENTS

The LIMS can keep track of instrumentation. Each equipment record can contain the following information: equipment name, equipment manufacturer's information, dates when equipment is received and placed in service, location of equipment, maintenance record, calibration dates and results.

If it is required that the instruments be able to directly parse their data to the LIMS, they will need to become part of the local area network. Each instrument will need to be identified by unique name or Internet Protocol (IP) address.

VI. COMPUTER NETWORK

The involvement of the Information Services (IS) department is critical for successful LIMS implementation. IS usually maintains the LANs, WANs, file servers, and computers. At the beginning of the project it must be determined which LIMS items the IS department be responsible for supporting and which will be the responsibility of the LIMS administrator.

All aspects of the LIMS should be discussed with the IS department, which should provide technical input and advice on infrastructural issues. Questions to be asked include: Are the LIMS database and operating system compatible with the operating system and databases already in place and supported by the IS department? Who will maintain the LIMS server, perform backups, apply service packs, maintain system security, and maintain software licenses? What will be the responsibilities of the IS department and of the LIMS administrator? These roles must be clearly defined at the start of the project.

VII. PERSONNEL

An additional LIMS function that should be evaluated by the LIMS implementation team is personnel maintenance. The LIMS can maintain documentation for personnel training, experience, and job descriptions.

VIII. LABORATORY STANDARD OPERATING PROCEDURES

The LIMS can store and manage the laboratory's standard operating procedures. It can document the history of test method revisions, including effective dates and retirement dates of the procedure. It can be used to maintain an inventory of the laboratory's standard operating procedures, each with its own unique document ID. The LIMS should also allow SOPs to be available online for analysts to view as they are working in the laboratory.

IX. QUALITY CONTROL

The LIMS should be able to handle quality control data and relate it to specific samples and analysis batches. Quality control standards, check standards, surrogates, matrix spikes, spike duplicates, blanks, and sample replicates should all be handled by the LIMS. It should be able to generate quality control charts and trends. Input from the quality control manager and analysts should be solicited so that the LIMS requirements will include all of the laboratory's quality control needs.

X. ACCOUNTING

The accounting department's needs must be solicited. Will the LIMS need to interface with accounting software? Should the LIMS have the ability to invoice customers? Many LIMS can track the costs associated with each test. Pricing can be programmed into the LIMS so that it is uniquely tailored for each customer or project. Information about each sample can be exported to accounting software for invoicing purposes.

XI. LEADERSHIP

Any single individual cannot determine a company's LIMS requirements. A LIMS implementation team must be composed of members from all parts of the organization that will be affected by the LIMS. A leader should be designated to provide focus to the team, preferably the designated LIMS administrator. The examination of laboratory workflow, paperwork, and LIMS questionnaires will help determine the functional requirements of the LIMS.

9
Critical Elements in Preparing a Request for Proposal

The purchase of a laboratory information management system (LIMS) normally involves a large investment of time and money. Statistics indicate that nearly 60% of LIMS fail. Reasons for failure include that the system does not work as advertised, the users are not comfortable with system, the vendor will not support and/or help maintain the system once it is installed, and the lab has not allocated enough resources to the project. Preparing a thorough request for proposal (RFP) can eliminate the first three reasons. The RFP should include information on the lab requesting the proposal, information on how the proposal will be evaluated, anticipated schedule, technical specifications, training, support, installation, costs, acceptance testing, and reference requirements.

An RFP should include at least four sections: information for proposers, technical specifications, questionnaire, and price schedule. A sample RFP is included in Appendix A.

I. INFORMATION FOR PROPOSERS

The RFP should include information about the type of laboratory requesting the proposal: a manufacturing lab, an environmental lab, or a pharmaceutical lab? What type of samples will be processed? How many samples does the lab process a year? Will the LIMS be for one lab or is a multisite implementation planned? It should include directions for how the proposal should be submitted, the due date, and the number of copies required. A contact name and number should be provided in case a vendor requires clarification.

The RFP should detail how the proposal will be evaluated. Is cost the determining factor? Will technical merit weigh more than company performance? The RFP should inform the vendors if a demonstration will be required from all vendors or if only a few vendors will be selected to give a demonstration. The RFP should note whether or not the demonstration would be scripted.

If known, an anticipated schedule for the project should be provided. The schedule should include the proposal due date, selection of short list of vendors, vendor demonstrations, selection of successful proposal, and purchase and implementation of the LIMS.

II. TECHNICAL SPECIFICATIONS

The technical specifications of the RFP should be detailed. The more information that can be provided to the vendor about the requirements of the laboratory, the more detailed and thorough the proposal will be. A detailed proposal will help the laboratory to make an informed decision concerning the purchase of a LIMS suited to the laboratory's particular needs. The technical specifications can be provided in an outline format. If possible, a table of compliance included in the RFP can be of great assistance to the vendor in preparing its proposal and to the laboratory in evaluating proposals.

A. System Configuration

A laboratory's RFP should provide information on the type of computer network it uses: Novell, Windows NT, or another topology? Does the laboratory want a client/server system or a desktop system? How are the computer clients configured? Is the laboratory planning on buying new hardware based on the requirements of the LIMS or will the LIMS need to run on the laboratory's current hardware configuration?

B. Database Management

Does the laboratory have a preferred database for the LIMS such as Oracle, Microsoft SQL server, or Access? How many concurrent users are expected to use the system? That determines the system's software licensing requirements. If the lab wishes to import or export data from the LIMS, it should be specified in the RFP. The vendor should be informed if data will need to be transferred from another LIMS or other database into the new LIMS.

C. Sample Tracking and Management

Sample tracking is a feature of every LIMS, but the way in which a sample is tracked by a LIMS is not the same for every vendor. The laboratory should inform vendors exactly how the laboratory wants a sample to be tracked. Should the LIMS be capable of differentiating between inhouse analyses and contract lab analyses? What data should be captured at sample login? Data may include sample collector, collection date and time, sample receiver, received date and time, sample location code and address, and field data.

Does the laboratory want the LIMS to login samples automatically based on a predefined schedule? Should the LIMS store static information about customers and sampling locations?

D. Sample Identification and Receiving

One feature that many LIMS programs have is the ability to prioritize workloads. When a sample is received into the laboratory, it can be assigned a priority code. It should be specified in the RFP if this is a desired feature. Samples can be logged into a LIMS in batches, individually, or both batches and individually. The RFP should describe how the laboratory wants to log in samples. Does the lab require the ability to associate comments with a sample? Does the condition of the sample need to be recorded during sample log in? Must the laboratory sample ID number follow a prescribed format? For many LIMS sample ID numbers follow the format of *yymmXXXX*, where *yy* is the year, *mm* is the month, and *XXXX* is a four-digit counter. Not all LIMS offer flexibility in the numbering scheme.

E. Test/Analysis Administration

Should the LIMS store calculation data about each test component? Does the laboratory require that a single test contain multiple parameters? How does the laboratory wish to enter analytical data? Results can be entered from one test performed on many samples, all results for many tests performed on one sample, or a combination of both ways. Does the laboratory need to enter text data as well as numerical data? Does data entry need to comply with good automated laboratory practice (GALP) regulations or electronic signature requirements? Will results need to be entered, validated, and approved prior to reports being generated, or should reports be able to be generated prior to approval? Many LIMS allow the formation of test groups where many tests can be grouped together. This can be a great benefit if a number of tests are routinely assigned to samples. The LIMS should be able to mimic or improve upon the laboratory's current data entry procedures.

F. Bench Sheets/Work Assignments

A LIMS can produce bench sheets. If a laboratory desires this feature, the RFP should specify what type of bench sheets are desired. Should bench sheets and work assignments be able to be produced based on test, instrument, or analyst? What information should be on the bench sheets? Should quality control samples be on the bench sheets? Only general requirements should be described: a detailed description for each bench sheet is not necessary.

G. Status Monitoring

Status monitoring is how the LIMS monitors the status of a sample through the sample's life cycle. Desired features often include the ability to update a sample's status automatically. What stages in a sample's life cycle does the laboratory wish to monitor: receipt, data entry, report generation, disposal? Does the laboratory want its customers to have the ability to access their data via the Internet or customer call up? This information should be included in the RFP.

H. Statistical Analysis and Quality Control

A common quality-control feature of many LIMS is the ability to associate sample analysis results with a set of quality-control data for specific analytical batches. The laboratory should include in its RFP the type of quality-control samples the laboratory analyzes: blanks, sample duplicates, matrix spikes, matrix spike duplicates, surrogates, standards, or field blanks.

Many LIMS can produce quality control charts but it should not be assumed that all LIMS provide this functionality. The RFP should specify whether or not this is required. The formatting of charts should be specified. Should the charts conform to EPA protocols? What type of statistical analysis should the LIMS be able to perform? Are warning limits and control limits desired?

I. On-Line Queries

A laboratory wants to be able to retrieve logically related data quickly and easily, in an interactive environment, without the need for detailed understanding of data storage and programming techniques. In what ways does the laboratory wish to retrieve data: by sample number, location code, test? Does the laboratory staff want the results displayed on the workstation screen, sent to a printer, saved as an ASCII file, or exported to another program?

J. Reporting

What types of reports does the laboratory require the LIMS to produce? Should the vendor provide preprogrammed reports such as sample backlog report, quality control (QC) reports, sample volume reports, and National Pollution Discharge Elimination System (NPDES) reports? Should the LIMS come with its own reporting software or is third-party software required? Does the laboratory require multiple reporting formats? This is a very important feature of the LIMS, if reporting is not flexible, then seek a system that is.

K. Instrument Interfaces

The RFP should specify if the laboratory wishes to interface analytical instrumentation with the LIMS. Specific instruments and associated data handling software should be detailed and sample output files provided.

L. Accounting Functions

Many LIMS systems provide basic accounting functions, such as costs per sample, quotation generation, and invoicing. Some LIMS interface with third-party software for accounting purposes. Accounting functions may be optional for some LIMS and the function can be turned off if desired, but other LIMS systems require a sample to be "invoiced" in order for a report to be generated.

M. Chemical Inventory

A chemical inventory module is available with most LIMS. The RFP should specify whether or not this should be included. Should the LIMS store vendor information and pricing? Should lab personnel be notified when to order supplies? Should the LIMS store MSDS information?

II. PRODUCT SUPPORT

Support options vary from vendor to vendor. Some vendors offer multiple support options while other vendors offer only one support option. The RFP should state the level of support that will be required. Should the LIMS vendor provide a toll-free number for support calls? Should it have a newsletter and user groups? What type of response is acceptable for a support call? What type of support quality is provided? Are the engineers technically capable, knowledgable in the software and able to provide solutions? Is a 24-hr turn-around time sufficient or does the lab require a 2-hr response time? Is 24-hr, 7 days a week support required or is support during normal business hours acceptable? How often should upgrade be provided? Should functional fixes be provided regardless of the support plan purchased by the laboratory?

What type of documentation will the laboratory require? Is documentation only on CD-ROM acceptable? Some LIMS vendors charge for hard copy documentation. How many user manuals are required? Usually one user manual will be provided at no charge.

III. TRAINING

The type and frequency of desired training should be included in the RFP. Should training be provided on-site or at vendor training centers? How many system administrators and how many end-users will require training? Will follow-up training be required? The RFP should specify that a course syllabus for all vendor-provided training be included in the vendor's proposal.

IV. FUNCTIONAL AND ACCEPTANCE TESTING

The RFP should require that the LIMS vendor provide a test plan and perform testing on the system after installation to demonstrate functionality and performance. The RFP should state the acceptance testing period that the laboratory will require before the LIMS receives final acceptance by the laboratory. Average acceptance periods range from 90 to 120 days. This allows the laboratory time to test the system and the vendor time to resolve any deficiencies. If the vendor cannot provide a functional LIMS according to the requirements of the RFP, the laboratory should not be obligated to pay for the LIMS.

V. PROPOSER INFORMATION QUESTIONNAIRE

The RFP should include an area where a vendor can provide additional information. A questionnaire can provide this function. Information might include: how long the company has been in business, address of the office supporting the LIMS, information on the firm's certifications (such as ISO 9001, Microsoft, Oracle, etc.), number of personnel dedicated to support and number of personnel dedicated to LIMS development, number of customers with the proposed LIMS system currently installed, and profile of the vendor's customer base. The vendor should be asked to provide a list of customers as references.

VI. COST SCHEDULE

The RFP should include a cost schedule to be completed by the vendor. Each item in the cost schedule should refer to specific sections of this group of work. Columns for unit price, quantity, and total price should be included.

10
LIMS Evaluations

A laboratory will begin to receive proposals from vendors for a Laboratory Information Management System (LIMS) in response to the request for proposal (RFP) developed by the laboratory. The proposals should be evaluated according to the vendor's ability to fulfill the functional requirements of the LIMS, the vendor's expertise with the type of laboratory requesting the proposal, costs, technical support, future system expandability, certifications/ qualifications, and references. If a vendor cannot meet the functional requirements of the laboratory or if the costs are prohibitive, the laboratory LIMS team will need to evaluate if the project should be abandoned, have a custom LIMS developed inhouse, or modify the functional requirements of the LIMS.

The LIMS proposal can serve as a contract between the laboratory and the LIMS vendor. The written proposals should be evaluated to develop a short list of vendors, usually three to five, that will be invited to give a scripted demonstration. If possible, all members of the LIMS team should evaluate the written proposal.

I. WRITTEN PROPOSAL
A. Hardware/Software Configurations

The vendor should indicate if their LIMS software would run on the laboratory's current computer system. If it cannot, technical specifications for required computer system components should be described in detail.

B. Database Management

The proposal should describe the database management tools available in the LIMS, the licensing requirements, and the ability to export or import data.

C. Sample Management and Tracking

The proposal should describe the tracking of the sample through the laboratory. What information is captured at log in? How can data be retrieved? What static information is stored for a sample location?

D. Sample Scheduling

The proposal should describe how samples could be scheduled. Can scheduled samples be pre-logged in? Can the sample information be modified? How will the system handle scheduled samples that are not collected?

E. Sample Identification

The proposal should describe how samples are identified in the LIMS.

F. Sample Receiving

The proposal should describe what information about a sample is captured at sample login? Can field data be captured? Are samples processed in batches or individually? Are holding times established at login?

G. Test Analysis

The proposal should describe how tests are defined in the LIMS and how they are assigned to samples. Can limits be placed on results: reporting limits, detection limits, or calibration limits? Can special results be handled? How are results entered?

H. Bench Sheets

The proposal should be evaluated to determine if it fulfills the laboratory's requirements for the production of bench sheets. Several examples of bench sheets produced by the LIMS should be included.

I. Status Monitoring

The methods for monitoring the status of a sample throughout its life cycle should be described. The status of the sample should be automatically updated based on events or transactions.

J. Quality Control and Statistical Analysis

The proposal should describe how quality control data are handled by the LIMS. Are the quality control functions an intrinsic part of the LIMS program, or is third-party software utilized for this? If third-party software is utilized, does the LIMS vendor support it? The production of control charts should be detailed and examples should be included in the proposal.

K. Instrument Interfaces

Can the LIMS interface with laboratory equipment? The proposal should describe how instruments are interfaced and if additional software or hardware is required.

L. Cost Functionality

The proposal should indicate if accounting features are included with the LIMS package. Can cost be assigned to individual tests or group of tests? If accounting features are included, must a sample be "invoiced" in order to archive a sample? Will the LIMS interface with other accounting software packages?

M. Support

The proposal should describe the various types of support options that the vendor offers. Good support for the LIMS is necessary. The vendor should have a good warranty, availability of software customization, on-site and toll-free telephone support, bug fixes and maintenance, software updates, user groups, and technical expertise. Most LIMS vendors will offer a variety of maintenance agreements that are renewable on an annual basis. Support may seem like an easy area in which to control costs, but it is not advisable. Good support is critical for successful LIMS implementation, especially for the first few years.

N. Documentation

The vendor should provide all documentation for the LIMS application. This should include installation instructions, an administrator's manual, an end-user's manual, and documentation on instrument interfaces. The source code should be provided or held in escrow in case the LIMS vendor should go out of business. The vendor should be able to supply documentation required for validation according to Good Automated Laboratory Practices.

O. Training

The proposal should include details on the training provided by the LIMS vendor. The vendor should be able to provide training on all items it supplies. Initial training should be on-site, with follow-up training available at the vendor's location. A course outline for all training courses should be included in the proposal.

P. Expenses

The proposal should itemize all costs. A review of the proposals will illustrate that the costs of the LIMS can vary widely from vendor to vendor. Similar systems can vary in cost by as much as $100,000. Mainframe, minicomputer, and client/server systems are more expensive than desktop systems. System costs are fairly easy to calculate and include the costs of a software package, maintenance agreements, upgrades to hardware if required, network cabling, network interface cards, validation, and training. Costs should be compared between vendors equally on an itemized basis. A vendor may include additional options that may inflate the total cost of the system compared to the total cost of another system.

Q. References

The proposal should include references provided by the LIMS vendor for the systems installed in laboratories similar to yours. The LIMS system should also be similar. If the proposal is for a client/server system, the references should be for a client/server system and not a desktop system. Client/server systems and desktop systems differ in complexity. Customer satisfaction with a desktop system may not be indicative of customer satisfaction with a client/server system. Features that work in a desktop system may not function well on a client/server system. All of the references should be thoroughly checked. The satisfaction of existing LIMS customers will be an important factor in the consideration of the vendor's proposal.

II. SCRIPTED DEMONSTRATION

The evaluation of the written proposals should yield a short list of vendors (three to five) that should be invited to give a scripted demonstration, in which the laboratory dictates what and how the LIMS vendor will demonstrate their product based on a script. Each vendor should be provided the same script. The purpose of the scripted demonstration is to give consistency between the vendors' demonstrations.

This script should be designed to test how a vendor's product will handle a typical sample in the laboratory seeking a LIMS. The chosen sample should encompass many of the quality control, quality assurance, and data reviews normally performed on data generated by the laboratory. Calculations should be included in the demonstration if the LIMS will be required to perform calculations. The LIMS should be able to generate a report in a format currently used by the laboratory. Vendors should not be allowed to deviate from the script unless they need to do so to continue processing the sample. The vendor should be prepared to discuss configuration and implementation recommendations.

Each member of the LIMS evaluation team should be present for each demonstration. An evaluation form can be developed to judge the demonstration. The final decision for the purchase of a LIMS should be based on how well the vendor's product processes the laboratory's sample through the various components of the script.

11
Enhancing Data Quality with LIMS

I. QUALITY CONTROL AND QUALITY ASSURANCE

Although quality control (QC) and quality assurance (QA) sound very similar, their meanings are not. QC refers to operational techniques and tests required to maintain and improve the quality of a product or service. QA refers to activities that monitor and evaluate the performance of QC procedures employed in the manufacture or products and services. Laboratory Information Management Systems (LIMS) make significant contributions to QA/QC through automatic specification checking, with user warnings, graph (control charts and trend analysis) and report generation, full audit trail, automatic calculations, automated sample tracking, on-line standard operating procedures (SOPs), maintenance of standards, instrument calibration, tracking training records, and much more.

Employees' training records can be tracked (with built in "lock-out"), to identify employees who are not trained or whose training has expired on a particular method or technique. This can also be done for an instrument on which the calibration has expired: it can be "locked out" from entering results into the LIMS. No further results can be entered until it has been recalibrated. These type of checks lead to a reduction in data entry errors and higher data quality, which translate to higher information quality on which better decisions can be based.

The Environmental Protection Agency's (EPA's) Good Automated Laboratory Practices (GALP) include requirements to ensure the following:

SOPs are followed.
Accuracy and correctness of all formulas, calculations, and algorithms.
Ability to audit all data entry and any modifications.
Change is a consistent and controlled process.

Integrity of all data entered into the LIMS (either manually or electronically).
A disaster recovery plan in place.

The requirements outlined above relate to many different areas of an
LIMS, including system, administrative, application, and network security,
data backup procedures, utilization of surge protectors, uninterruptible power
supply, and system validation. Data entry is a critical step in the data validation
process and has been singled out by the EPA's GALP as an area that should
be carefully examined. LIMS help significantly in this role by restricting users'
entry to certain departments and tests, with a user name and password. In
addition, there are multiple layers of security (password), from the LIMS, to
the administrator, to the database engine, and also network security. Typically,
users need to supply a username and password to log onto the network. There
is also built-in limit checking. Many laboratories employ bar codes, in addition
to limiting users to selections from pick lists, providing user warnings and
prompts, and data validation using specifications and autoreporting. Some of
these items are described in greater detail below.

II. SYSTEM VALIDATION

Validation is typically required if the LIMS will operate in a regulated environment. This is becoming more and more the case as regulated environments
expand across all industries, and is discussed in greater detail in the next chapter (Chap. 12). Validation entails proving or verifying that the LIMS software
is operating in a predictable and correct manner for its intended purpose. Vendors may supply validation scripts, which allow end-users to run through a
series of functions, input a known value, and make sure that the output is what
was expected, based on the input. Validation is a very tedious process and
differs from testing; testing evaluates if the software meets the design specifications. Vendors typically work with clients to perform system customization,
and they can assist clients in the validation process by delivering a thoroughly
tested and documented system.

III. STEPS TO LIMIT DATA ENTRY ERRORS

A. Data Entry Restriction

A reliable LIMS function is checking information entered into the LIMS to
ensure that it is of the correct format and field size (e.g., a text character should

not be allowed in a numeric field). There can also be automatic limit checking (e.g., the LIMS will not allow users to enter a pH of 25) and there can be validity checking (e.g., you cannot analyze a sample before the manufacture date).

B. Double Data Entry Screens

This feature asks either the same data entry person or a different one to re-enter in either specified fields or all fields to see how the two entry data screens compare. If information is keyed in differently from the first entry typically a highlighted field or fields indicate the differences. This feature is frequently employed by those in the clinical fields and typically those involved in patient demographic entry.

C. Range or Limit Checking

Built-in upper and lower warning limits can automatically alert a user that data is outside an upper or lower warning range or a hard limit (e.g., a pH of 18). These ranges are usually established by the user, for test ranges, client specifications, or user definition.

D. Limit to List

The utilization of pick lists greatly reduces spelling errors and forces consistency in test names and other common pull-down list items. For example, for a particular matrix, it will only display tests that can be performed on that particular matrix. The user's choices will be limited to the items on the pull-down list, and a "hot" look-up is often employed so that the user need just begin typing the item and the list will jump to that item. Pick lists are extremely useful for avoiding data entry errors, especially typographical errors. Another feature that is also helpful, but not as critical as limit to list, is the ability to prompt the user to perform a function. For example, after a user audits a result record, and before they close the screen, they receive a message reading "You must provide a reason for the change before you can close." Another feature that helps to decrease data entry errors is the use of bar-codes, which can contain information about the sample avoids transcription errors and saves keystrokes in entering information.

E. Automatic Calculations

Computers are ideal for performing routine calculations. When combined with the sample tracking and data entry functions of a LIMS they are ideal at calculating turnaround times, due dates, percentage recovery, and any other required calculation. If the calculation is properly set up, tested, and verified, this also ensures accurate calculations and decreases the potential for transcription errors.

F. Automatic Reporting

Perhaps one of the greatest time savers for the laboratory is the autoreporting function in many LIMS. Today autoreporting means more than a simple print-out: it may involve automatic e-mail notification for products that are out of specification, or automatic faxing of results that have been approved and validated. In addition to common reports, such as production, backlog, QC, and certificates of analysis, graphs can also be automatically generated.

G. LIMS and Data Validation

LIMS assist in data validation in several ways. Several have been discussed previously, such as limits on the data entry format, and automatic limit checking. Others include a double entry screen where the same or a different user enters in the same data to ensure accuracy.

H. Decreased Turnaround Time

Many of the quality assurance features of a LIMS that have been described above also serve to increase sample throughput:

> Bar-coding (automatic data entry)
> Instrument integration
> Autoreporting (via printer, fax, or e-mail)
> Automatic calculations
> Rapid data retrieval
> Data validation (upon entry, limit checking)

Not only are transcription errors virtually eliminated and data quality improved, but the laboratory throughput also increases significantly because analysts are spending less time interacting with paperwork and performing mundane tasks. The chemists will have more time to analyze additional tests, perform method development, and other more challenging functions.

IV. SECURITY

A major advantage of LIMS over a manual paper-based system is the multiple levels of security available to protect data integrity and allow laboratory personnel access to specific sections of the LIMS via passwords and to have a log of permissions and access.

A. System Security

The task of assigning permission to access specific functionality to users of the LIMS usually lies with the database administrator. The database administrator has many responsibilities, including maintaining the static tables of the LIMS, assigning usernames and passwords, performing regular backups, and modifying report templates. If proper security is lacking in the LIMS, users may either accidentally or maliciously modify data. The more critical the information held in the LIMS, the tighter the security required.

B. LIMS Application Security

A good LIMS will have multiple levels of security: view-only status; permissions for specific departments, but not others; view and approve permissions; or view, approve, and validate. The fact that users must log into most LIMS with a user name and password identifies the time the user was logged onto the system and what results he or she entered. Like any system, it is only as secure as the users. If users share passwords, never logoff, or logon using co-workers' passwords, there is no good way to track changes or data entries. Stricter security procedures would then need to be implemented, including frequent changes to passwords and periodic reviews of user logs.

C. Network Security

Networked environments provide another level of security and the network administrator typically assigns permissions to individuals or groups to access the network. For example, network security should provide security for each user, directory, and file. Network security rights typically include, read, write, delete, create new files, create new folders, and search capabilities.

The examples provided here give the reader a good idea of how a LIMS can offer a laboratory increased assurance that high-quality data are generated.

12
LIMS Validation

Validation of a laboratory information management system (LIMS) is becoming an increasingly important issue for many laboratories. A host of guidelines such as Good Laboratory Practice (GLP), Good Automated Laboratory Practice (GALP), and Good Clinical Practice (GCP), and regulations such as Good Manufacturing Practice (GMPs) and quality system regulation have been issued steadily over the past several years. The objective of the validation process is to ensure that a system does what it purports to do and will continue to do so. Validation not only satisfies regulatory requirements but also is a good tool for organizations to use so that they can feel confident that the LIMS performs the way they expect it to.

There is no single, standard way to plan and implement a validation process. This chapter does not recommend any specific life-cycle model or specific validation technique or method, but it does recommend that validation activities be conducted throughout the entire LIMS life-cycle. The process should start with the planning of the requirements of the new system and continue through specification, testing, implementation, operation, and retirement of a system. For example, as specifications for new system are developed, the validation process will later test and examine whether the specified system will meet the defined requirements.

I. PRINCIPLES OF SOFTWARE VALIDATION

There are 10 general validation principles considered applicable to an LIMS.

1. Timing. Validation is not a one-time event. It should begin when planning and input begin. Validation does not end until the product is no longer used.

2. Management. Proper validation of an LIMS includes the planning, execution, analysis, and documentation of appropriate validation activities and tasks.
3. Plans. Established design and development plans should include a specific plan for how the software validation process will be controlled and executed.
4. Procedures. Validation procedures should be developed. The validation process should be conducted according to the established procedures.
5. Requirements. To validate a LIMS, there must be predetermined and documented requirements. If a request for proposal was thoroughly developed, it will contain the requirements necessary for validation.
6. Testing. Verification includes static and dynamic techniques. Static techniques include paper/document reviews; dynamic techniques include physical testing to demonstrate the system's run time behavior in response to selected inputs and conditions. Dynamic analysis alone may be insufficient to show that the system is fully functional and free of avoidable defects. Static techniques are used to offset limitations of dynamic analysis. Inspections, analyses, walk-throughs, and design reviews may be more effective in finding, correcting, and preventing problems at an earlier stage of the development process.
7. Partial validation. A system cannot be partially validated. When a change is made to the system, the validation status of the entire system should be addressed.
8. Amount of effort. The magnitude of the validation effort should be commensurate with the risk associated with dependence on critical function. The larger the project and staff involved, the greater the need for formal communication, more extensive written procedures, and management control of the process.
9. Independence. Validation activities should be conducted using the basic quality assurance concept of "independence of review."
10. Real world. It is fully recognized that LIMS are designed, developed, validated, and regulated in a real-world environment. Environments and risks cover a wide spectrum and that each time a validation principle is used, the implementation may be different.

During the concept phase of a LIMS, one or more staff members may be assigned to consider and document the project, its purpose, anticipated users, intended use environments, system needs, and the anticipated role of the software. The concept may include basic system elements, sequence of operations, constraints and risks associated with development, and performance requirements for the system. The following is a list of preliminary considerations:

> How will off-the-shelf (OTS) software be validated?
> What are the risks and benefits of OTS vs. contracted or in-house-developed software?
> What information is available from the OTS vendor to help in validating use of the software?
> Will the OTS vendor allow auditing of their validation activities?
> Who will control the source code and documentation?

II. LIFE-CYCLE ACTIVITIES

Activities in a typical LIMS life-cycle include:

> Management/project initiation phase
> Requirements phase
> Design phase
> Implementation phase
> Integration and test phase
> Installation and acceptance phase
> Operation and support phase

For each of the life-cycle activities, certain validation tasks are performed.

A. Management/Project Initiation Phase

During design and development planning, a validation plan is developed to identify required validation tasks, procedures for reporting anomalies and their resolution, resources needed for validation, and management review requirements. The validation plan should include:

> Specific validation tasks for each life-cycle activity
> Methods and procedures for each validation task
> Criteria for acceptance of completion of each validation task
> Inputs for each validation task
> Outputs from each validation task
> Criteria for defining and documenting outputs
> Roles, resources, and responsibilities for each validation task
> Risks and assumptions

21 CFR 820 requires that management identify and provide the appropriate validation environment and resources. Each validation task will require personnel as well as physical resources. The validation plan should identify the personnel, facility, and equipment resources for each validation task. Procedures should be created for the review and approval of validation results, including the responsible organizational elements for such reviews and approvals. An outline of a test script for validation is provided in Table 1.

B.   Requirements Definition Phase

A LIMS requirement specification document should be created with a written definition of the software functions to be performed. It is not possible to validate a LIMS without predetermined and documented requirements. Typical requirements specify the following:

> All inputs the system will receive
> All outputs the system will produce
> All functions the system will perform
> All performance requirements the system will meet (data throughput, reliability, timing, etc.)

Table 1   Outline of a Test Script for Validation

1. Test script identifier: Unique identifier of the test script.
2. Purpose: Describe the feature to be tested.
3. Special Requirements
4. Test Procedure Steps
   Describe procedures to be carried out.
   Exact steps to be carried out.
   Cross-reference any documents, such as a user manual.
5. Test Log
   Describe expected results.
   Write observed results.
6. Unexpected events: Describe any unexpected events, such as test not working.
7. Resolution of unexpected event: Describe any steps taken to resolve unexpected events
8. Pass/Fail criteria
   State the pass/fail criteria for the test.
   Compare expected and observed results.
   Does the test pass or fail?
9. Sign-off by tester and peer reviewer.

Definition of all internal, external, and user interfaces

What constitutes errors and how errors should be handled

Internal operating environment for the software (hardware platform, operating system, etc.)

All ranges, limits, defaults, and specific values the software will accept

The software requirements interface analysis should be conducted comparing the software requirements to hardware, user, operator and software interface requirements for accuracy, completeness, consistency, correctness, and clarity to ensure that there are no inconsistencies. During the requirements definition phase, the validation activities should ensure that the requirements are testable.

C. Design Phase

During the design phase, specifications should describe the logical structure of the LIMS, parameters to be measured or recorded, information flow, logical processing steps, data structures, control logic, error and alarm messages, security measures, and predetermined criteria for acceptance. The design phase also describes any supporting software such as operating systems, drivers, or other applications that may be required. Special hardware that will be needed, communication links among internal LIMS software modules, links with supporting software, links with hardware, and any other constraints not previously noted are to be identified.

Design specifications may include:

Data flow diagrams

Program structure diagrams

Control flow diagrams

Interface/program diagrams

Data and control elements definitions

Module definitions to

Module interaction diagrams

Validation activities that occur during this phase have several purposes. Design evaluations are conducted to determine if the design is complete, correct, feasible, and maintainable. Appropriate consideration of system and software architecture at the design phase can reduce the magnitude of future validation efforts when software changes are needed. At the end of the design phase, a formal design review should be conducted to verify that the design is correct, consistent, complete, accurate, and testable before moving to implement the design. Typical validation tasks for this phase include:

    Test design generation
    Software design evaluation
    Design interface analysis
    Module test plan generation
    Integration test plan generation
    Finalization of acceptance test plan

D. Implementation Phase

In this phase, detailed design specifications are implemented as source code. For commercial LIMS products, the implementation phase is carried out by the vendors' programming departments. For in-house-designed LIMS, the implementation phase is conducted by in-house programmers or contracted programmers.

Decisions on the selection of programming languages and software building tools usually occur in the requirements or design phases of development. Some compilers offer optional error checking commands and levels. For validation purposes, if the most rigorous level of error checking is not used for translation of the source code, justification for the use of less rigorous error checking should be documented. There should also be documentation of the compilation process and its outcome.

The source code should be evaluated to verify its compliance with specified coding standards, which should include conventions for clarity, style, complexity, management, and commenting. Code comments should provide useful and descriptive information for a module including expected inputs and outputs, referenced variables, expected data types, and operations to be performed. Source code evaluations should be extended to the verification of internal interfaces between modules and compliance with design specifications. Appropriate documentation of source code evaluations should be maintained as part of the validation information.

Static and dynamic, formal and informal testing methods may be employed during the implementation phase. Static methods include the code evaluations as described. Dynamic analysis begins after a source code module has passed the necessary static code evaluations. Dynamic testing may be informal as a programmer refines a module's code to conform to specifications and become more formal with documentation when module testing begins.

A source code traceability analysis should be conducted. This is an important tool for verifying that all code is linked to established specifications and test procedures. The analysis will document verification that each element of the software design specification has been implemented in code, that modules and functions implemented in code can be traced back to the design specification, and that tests for modules and specification can be traced back to the design specification.

Typical validation tasks for the implementation phase include:

Traceability analysis
    Source code to design specification
    Test cases to source code and design specifications
Source code and source code documentation
Source code interface analysis
Finalized test procedure and test case generation (module, integration, system, and acceptance)

E. Integration and Test Phase

During the integration and test phase, the tested units from the previous phases are integrated into subsystems and then the final system. Testing objectives include the demonstration of compliance with all software specifications and the production of evidence providing confidence that errors have been identified and removed. A software testing strategy design to find and correct defects will produce far different results than the strategy designed to prove that the software works correctly. A complete testing program uses both strategies.

Integration is the process in which the individual modules making up a system are combined until the complete program has been assembled. Integration methods range from nonincremental integration to any of the methods employed for incremental integration. Nonincremental integration is often used for small programs; incremental integration methods are typically used for large programs. The properties of the program being assembled dictate the chosen methods of integration.

Test plans should be created during the prior software development phase. They should identify the test schedules, test environments, resources (people, tools, etc.), methodologies, cases (inputs, procedures, outputs, expected results), documentation, and reporting criteria. The test plans should be linked to each of the specification phases: requirements, design, implementation, and others. Individual test cases should be associative with particular specification elements and each test case should include a predetermined, explicit, and measurable expected result derived from specification documents in order to identify objective success/failure criteria.

Test plans should identify the extent of testing as well as predetermined acceptance criteria. The magnitude of testing should be linked to criticality and reliability. Each externally visible function and each internal function should be tested at least once. Detailed written procedures and checklists are often used to facilitate consistent application of intended testing activities.

Testing personnel should be independent of programming personnel. Testing personnel should have adequate knowledge of the software application's subject matter and software/programming concerns related to testing.

The methodologies used to identify test cases should allow for a thorough examination of the LIMS application and should include structural and functional testing. Structural testing examines the program's data structures: configuration tables, control and procedural logic, and nonfunctioning code. Structural testing should ensure that the program's statements and decisions are fully exercised and examined. Testing should be done at the module, integration, and system levels of testing.

Functional testing should expose program behavior in response to the normal case and in response to worst-case conditions. Functional testing will demonstrate program behavior to input and output domains, responses to invalid, unexpected, and special inputs. Functional testing should be conducted at the module, integration, and system levels of testing.

Module testing focuses on the examination of subprogram functionality and ensures that functionality not visible at the system level is examined. It should be done prior to the integration phase to confirm that each module meets specifications and that only quality modules are integrated into the finished LIMS software.

Integration-level testing focuses on the transfer of data and control across a program's internal and external interfaces. External interfaces are those with other software programs including operating systems software, system hardware, statistical software, other business software, and end-users. Internal interfaces are those among the program's various modules. When a program is built using incremental integration methods, sufficient testing should be performed to ensure that the addition of new modules has not changed the behavior of existing ones.

System-level testing demonstrates that all specified functionality exists and that the software is dependable. Test cases designed to address concerns such as robustness, stress, security, recovery, usability, and other issues should not be used to verify the software's reliability. System-level testing should demonstrate the LIMS's behavior in the intended operating environment. Test plans should identify the controls used to ensure that the internal level of testing is achieved and that proper documentation is prepared when planned system testing is conducted at sites not directly controlled by the software developer, such as customer locations.

Test results should be documented so as to permit objective pass/fail decisions to be reached. The results should be suitable for review and decision-making subsequent to running the test. Test results should be suitable for use in any subsequent regression testing. All errors detected during testing should be logged, classified, reviewed, and resolved prior to release of the software. The test reports should comply with the requirements of the test plans.

Typical validation tasks for the integration and testing phase include:

Traceability analysis
    Module tests to detailed design
    Integration tests to high-level design
    System tests to software requirements
Test evaluation
Error evaluation/resolution
Module test execution
Integration test execution
System test execution
Test results evaluation
Final test report

F. Installation and Acceptance Testing Phase

Installation testing is an essential part of validation for a LIMS. Terms such as beta test, site validation, user acceptance test, and installation verification have all been used to describe installation testing. Installation testing is any testing conducted at a user's site with the actual hardware and other software that will be part of the installed LIMS configuration. The testing is accomplished through either actual or simulated use of the software being tested within the environment in which it is intended to function. If the computers for the LIMS are placed on a network, a validation procedure for proper network installation and connection will need to be established. All modems, printers, fax machines, and instruments for data acquisition will need to be tested for their proper integration into the LIMS.

Installation testing should follow a predefined plan, with a formal summary of testing and a record of formal acceptance. Keep all documented evidence of testing procedures, test input data, and test results. There should be evidence that hardware and software are installed and configured as specified. The testing phase should continue for a sufficient amount of time to allow the system to encounter a wide spectrum of conditions and events so as to detect any latent faults not apparent during normal activities.

In addition to an evaluation of the ability of the LIMS to perform its intended functions properly there should be an evaluation of the ability of its users to understand and correctly interface with it. Users should be able to perform the intended operations and respond in an appropriate and timely manner to all alarms, warnings, errors, or other situations demanding action.

Several areas require special attention during an LIMS validation study. All procedures for data entry and receipt of data, either electronic or manual, must be formal and implemented to ensure consistent execution. Data ranges should be determined, as well as system edits activated to limit the introduction of erroneous data. Each input field should be identified and the allowable input defined. Testing should be conducted to demonstrate accurate processing of valid or "good" data and the rejection of invalid or "bad" data.

If the LIMS performs calculations, evidence of the reliability of the formula/algorithm must be documented. Documentation often takes the form of published algorithms. Most hard coded calculations or pre-existing formulas can be traced to a published source.

All reports generated by a LIMS must conform to be established criteria for format content, and accuracy. Pre-established criteria may be part of an LIMS request for proposal (RFP). The specification document for each system output should describe the contents and format of the report. For a user-defined report, the report detail should be defined and configuration management documented.

Validation activities during this phase normally include:

Test, analysis, and review that the system meets requirements
Locate, correct, and retest nonconformances
Acceptance testing and reviews

G. Operation, Maintenance, and Support Phase

After a system is validated and becomes operational, changes will usually occur during its operational lifetime that may have an impact on validation status. The issues that may arise include:

Revalidation criteria
Configuration management

Change control
Audit trails
Standard operating procedures (SOPs)
Operational and maintenance records
Error logging and resolution

Any change to the LIMS should trigger consideration or revalidation of the system. There may be instances in which no revalidation would be necessary after a change. One way to evaluate a change is to review the impact the change would make to the data's accuracy, security, and integrity. This will allow for targeting of revalidation efforts.

Examples of changes to a system include hardware maintenance and upgrade, upgrade of the operating system, and evolution of the LIMS application overtime. Configuration management is a set of procedures to ensure adequate identification, control, visibility, and security of any changes made to hardware, firmware, network, program source code, or any specialized equipment associated with the LIMS.

The process of configuration management is simple. The initial system configuration is thoroughly documented. All components of the system should be listed: all the release numbers and serial numbers of the application software and the operating system. The components that make up the hardware, such as disks, memory, type of central processing unit, and any peripherals, as well as any documentation used with the system should be listed as well. As modifications to the system configuration are made, that information should be recorded in the configuration log.

Change control defines responsibilities and documents the process of change. Change can include the resolution of system bugs and errors. The impact of the change to the LIMS needs to be evaluated. Items to consider include time required to make the change, cost of the change, resources to make the change, and benefits of the change. The effects of the change should be documented. Operational logs will need to be updated to reflect any changes.

Validation activities are conducted throughout the entire LIMS lifespan. It starts with the requirements phase and continues through the operational phase. Even when a system is retired, it must be noted in the operational logs for future reference. Proper validation of an LIMS system will allow a laboratory to comply with regulations and also provide thorough documentation on the system that will be needed to help troubleshoot any problems that may occur.

References

Accelerated Technology Laboratories, Inc. Sample Master Pro LIMS user manual, 2001.

Albu, ML. ISO 9000 and the analytical laboratory. Am Lab September, 1996.

Andleigh PK, Gretzinger MR. Distributed Object–Oriented Data Systems Design. Upper Saddle River, NJ: Prentice Hall, 1992.

ASTM. Standard Guide for Laboratory Information Management Systems. December 1993, E 1578–93.

Atzeni P, De Antonellis V. Relational Database Theory. Menlo Park, CA: Benjamin Cummings, 1993.

Avery G, McGee, C, Falk S. Implementing an LIMS: a how-to guide. Anal Chem 72: 57A–62A, 2000.

Center for Biologics Evaluation and Research. Guidance for Industry: Computerized Systems Used in Clinical Trials. Washington, DC: FDA, April, 1999.

Cianfrani, CA, Tsiakals JJ, and West JE. ISO 9001: 2000 Explained Second Edition. 2000 ASQ Quality Press Publications.

Clark DL, Akiyoshi E, Oldewage L, Denger LB, Thompson KA. Design, selection and implementation of a laboratory information management system into a water quality laboratory—the do's and don'ts. Proc Water Qual Technol Conf American Water Works Association, 1997.

Codd EF. The Relational Model for Database Management, version 2. San Francisco: Adison-Wesley, 1990.

Date CJ. An Introduction to Database System, 6th ed. San Francisco: Addison-Wesley, 1995.

Schoeny DE, Rollheiser JJ. The Automated Analytical Laboratory: Introduction of a New Approach to Laboratory Robotics, American Laboratory September 1991, p. 42.

Dicorpo J, Paul S. Automating network backup and restore. Unisphere September/October, 1999.

EPA. Toxic Substances Control Act (TSCA); Good Laboratory Practices Standards. 40 Code of Federal Regulations (CFR) part 792, August, 1989.

EPA. Federal Insecticide, Fungicide and Rodenticide; Good Laboratory Practices Standards. 40 Code of Federal Regulations (CFR) part 160, August, 1989.

EPA. Manual for the Certification of Laboratories Analyzing Drinking Water. 4th ed. Washington, DC, 1996.

EPA. Good Automated Laboratory Practices, Recommendations for Ensuring Data Integrity in Automated Laboratory Operations with Implementation Guidance. Washington, DC, December, 1990.

EPA. Good Automated Laboratory Practices: Principle and Guidance to Regulations for Ensuring Data Integrity in Automated Laboratory Operations. Washington, DC, August, 1995.

FDA. Electronic Records, Electronic Signatures. 21 Code of Federal Regulations (CFR) part 11, April, 1999.

FDA. Good Manufacturing Practice Standards. 21 Code of Federal Regulations (CFR) parts 210, 211 and 820, 1993.

FDA. Current Good Manufacturing Practices for Finished Pharmaceuticals. 21 Code of Federal Regulations (CFR) part 211, 1996.

FDA. Quality System Regulation. 21 Code of Federal Regulations part 820, 1996.

Feghhi J, Williams P. Digital Certificates: Applied Internet Security. Reading, MA: Addison-Wesley, 1998.

Ford W, Baum MS. Secure Electronic Commerce: Building the Infrastructure for Digital Signatures and Encryption. Upper Saddle River, NJ: Prentice Hall, 1997.

Garner WY, Barge MS, Ussary JP. Good Laboratory Practice Standards. Washington DC: American Chemical Society, 1992.

Gillespie H. ISO 9000 in scientific computing. Sci Comput Automat February, 1994.

Gillespie H. Lab accreditation is all over the map. Today's Chem Work July, 1988.

Gillespie H. The electronic John Hancock. Today's Chem Work June, 1997.

Golden JH. Understanding the human dimension in LIMS projects. LIMS Lett December, 1998.

Grant GL. Understanding Digital Signatures: Establishing Trust Over the Internet and Other Networks (CommerceNet). New York: McGraw Hill, 1997.

Hertz CD, Bardsley AR, Kurisko PA, Senft MB. On Golden LIMS: Reflections on a Laboratory Information Management System. Proc Water Qual Technol Conf American Water Works Association, 1997.

Hines E. LIMS: the Laboratory's right arm. Pharm Formul Qual May/June: 37–43, 1999.

Hinton MD. Laboratory Information Management Systems—Development and Implementation for a Quality Assurance Laboratory. New York: Marcel Dekker, Inc., 1995.

Hoyle D. 1998. ISO 9000 Quality Systems Handbook, third edition. Butterworth-Heineman.

Huber L. Validation of Computerized Analytical Laboratories. Englewood, CO: Interpharm Press, 1995.

Huber L. Implementing 21 CFR Part II in analytical laboratories: part 1, overview and requirements. BioPharm 12(11):28–34.

ISO 25 emerging as lab accreditation standard. LIMS Lett February, 1998. Johnson T, Hanson R. Training is the key to success. Sci Comput Automat April, 1999.

Karbo M. The PC technology guide. www.PCTechGuide.com, 1999.

Lee GW. Bar coding and automatic identification technologies; an introduction to improving productivity and accuracy. In: The Auto ID Book. 3rd ed. Informatics, 1998.

McArthur D. Client/server issues for scientific information systems. Sci Comput Automat February, 1997.

McDowell RD. Operational measures to ensure continued validation of computerized systems in regulated or accredited laboratories. Lab Automat Info Manage 31, 1995.

McDowall RD. Biometrics: the password you'll never forget. LCGC Eur 13(10):736, 2000.

National Institute for Science and Technology. Software Verification and Validation: Its Role in Computer Assurance and Its Relationship with Software Project Management Standards. 1996.

Martinott RT. Laboratory Information Management Systems. Today's Chemist at Work, April 1992. P. SR3.

Paszko C, Miller T, Vranken R. Plugging Into LIMS: Evolution and Advances in the Age of the PC's; How Personal Computers Have Revolutionized Laboratory Automation and Data Management. Environ Test. & Analysis, September 22–24, 1998.

Paszko C, Pugsley C. Considerations in selecting a Laboratory Information Management System (LIMS) American Laboratory, September. 2000.

Paszko C. Extending LIMS Functions Over the Internet. Inside Laboratory Management. April 1998. 19–20.

PC Magazine. Evolution of storage devices. May 25, 1999.

Preparing and planning a new LIMS for an environmental laboratory. Sci Comput Automat May:37–39, 1997.

Preston C. Storage management technology. Windows NT Systems July, 1999.

Roberts J. Choosing a LIMS: a crucial decision for the laboratory. Sci Comput Automat October:54–55, 1999.

Roman S. Access Database Design and Programming Sebastopol, CA: O'Reilly & Associates, 1997.

Sanin L, Chen R. Client/server programming with Access and SQL Server. Ventura Communications Group, 1997.

Schmidt I. Client/server ODBMSs. Software Dev October:45–51, 1993.

Segalstad S. LIMS validation requires a systematic approach. Sci Comput Automat May 1997. 33–34.

Simovici D, Tenney R. Relational Database Systems. San Diego, CA: Academic Press, 1995.

Tetzlaff RF. GMP documentation requirements for automated systems: part I. Pharm
    Technol April: 60–72, 1992.
Tetzlaff RF. GMP documentation requirements for automated systems: part II. Pharm
    Technol April: 60–72, 1992
Ullman J. Principles of Database and Knowledge-Base Systems, vol 1: Classical Data-
    base Systems. Washington, DC: IEEE Computer Science Press, 1988.
Weber J. Bringing LIMS to the desktop. Am Lab July, 1994.
Weinberg, Spelton & Sax, Inc. GALP Regulatory Handbook, Lewis Publishers, 1994.
Windows NT Professional. 5(9), 2000.

Glossary

SPECIAL TERMS IN CLIENT/SERVER LIMS AND AUTOMATION

Acceptance testing—Formal testing conducted to determine whether a system satisfies its acceptance criteria and to enable the customer to determine whether to accept the system.

Accuracy—The closeness of agreement between the measured value and the true or accepted reference value. It is a measurement of the bias of a method.

ACIL—American Council of Independent Laboratories.

Address—An identification (number, name, or label) that uniquely identifies a computer register, memory location, or storage device.

Algorithm—A computational procedure containing a finite sequence of steps. A set of rules that specify a sequence of actions to be taken to solve a problem.

AIA—Analytical Instruments Association.

AIM (Automatic Identification Manufacturers)—An organization supported by manufacturers and suppliers of automatic identification products and services.

Analog signal—Any form of data transmission in which the pneumatic, mechanical, or electrical energy signal varies continuously in direct proportion to the intensity of the physical quantity, property, or condition represented.

ANSI (American National Standards Institute)—A nongovernmental organization responsible for the development of standards such as manufacturing.

Application software—A program that performs a task or process specific to a particular end-user's need, or solves a particular problem. Enterprise applications are typically large-scale business systems that organizations use to manage their operations.

Architecture—A structured set of protocols that implements a system's functions. Network architecture defines the functions, formats, interfaces, and protocols required for end-users to exchange information.

Artificial intelligence—The concept that computers can be programmed to assume capabilities such as learning, reasoning, adaptation, and self-correction.

ASCII (American Standard Code for Information Interchange)—A binary character code used to represent a character in a computer. It consists of 128 seven-bit codes for upper and lower case letters, numbers, punctuation, and special communication control characters.

Aspect ratio—The ratio of bar height to symbol length.

ASQC—American Society for Quality Control.

Association—A relationship between entities or data elements in a logical data model.

Assurance—A measure of confidence that the security features and architecture of an LIMS accurately mediate and enforce the security policy.

ASTM—American Society for Testing and Materials.

Attribute—A piece of information that represents a single property of an entity.

Audit—A qualitative and quantitative evaluation of the documentation and procedures associated with the LIMS to verify that resulting LIMS raw data are of acceptable quality.

Audit trail—A system that tracks the changes to a database.

Autodiscrimination—Capability of reading and decoding more than one bar code symbology in a single piece of equipment.

Automation—1. The conversion to and implementation of procedures, processes, or equipment by automated means. 2. The entire field of investigation, design, development, application, and methods to render or make manual processes or operations partially or fully automatic. 3. A system or operation that automatically compensates and adjusts itself to complete a task, or series of tasks, to predefined parameters without human interaction.

Background—Area surrounding a bar code, including quiet zone and spaces.

Backup—The process used to copy software, especially in regard to a database to another area or method of storage to protect the information.

Bar—The darker (black) element of a bar code symbol.

Bar code—An array of parallel bars and spaces encoding information. Also see *Symbol*.

Bar coding—An automatic identification technology that encodes data in a printed pattern of varying-width bars and spaces in accordance with predetermined rules.

Bar length—The bar dimension perpendicular to the bar width.

Bar width—The thickness of a bar measured from the edge closest to the symbol start character to the trailing edge of the same bar.

BASIC (Beginner's All-Purpose Symbolic Instruction Code)—A simple programming language widely used for personal computers.

Batch—A prepared quantity of material during one process operation.

Benchmark—A fixed point of reference or a standard for comparison; an outstanding example that is appropriate for use as a model.

Bias—The systematic error between the measured value and the true or accepted value.

Binary code—A representation of information using a sequence of zeros and ones; the basis for calculations in all digital computers.

Bit—A digit in binary code: 0 or 1.

Blank—An element of the analytical measurement process distinguished by the absence of the component to be measured. 1. Field blank—A blank used to measure the environmental contamination of a sample in the field. 2. Method blank—A blank composed of reagents, apparatus, and testing materials used in the test method.

Buffer—A storage area that temporarily holds data transmitted between a peripheral device and the central processor to allow for differences in working speeds.

Bus—A high-speed pathway shared by signals from several components of a computer.

Byte—1. A fixed number of bits, often corresponding to a single character and processed as a unit. 2. A collection of eight bits capable of representing an alphanumeric or special character.

Cache memory—A high-speed, buffer-type memory filled at medium speed from the main memory.

Calibration—The process of determining and/or adjusting an entity or device to meet or match a predetermined set of conditions or standards.

CAS—The Chemical Abstract Service number.

CCD (Charge Coupled Device)—A bar code scanner that senses the light and dark areas of a symbol.

CD-ROM (Compact Disk–Read Only Memory)—Systems that use digitized multimedia signals to recreate text, video, and graphics.

Certificate of analysis—A document that reports and certifies the test results of a product.

Certificate of conformance—A document confirming that a product or service meets the required specifications, regulations, or contractual agreements.

CGMP (current good manufacturing practice)—Updated versions of GMPs released by the federal government.

Chain of custody—Represents the record of a sample, including its collection, preservation, transportation, transfers, analysis, and final disposal.

Change Control—Management and implementation methodologies associated with increasing or correcting system capabilities, a partial system redesign, or determining software obsolescence.

Characteristic—A distinguishing property of a sample or population that can be measured or counted.

Check digit—A character included within a symbol used to perform a mathematical check to ensure the accuracy of the scanned data.

Check standard—A stable inhouse standard that is remeasured periodically to determine if a measurement process is in control.

CIM—(computer integrated manufacturing).

CIMS—(corporate information management system).

Class—A group of objects with common properties.

Client—A user's workstation in a client/server architecture, typically a PC or workstation. The client serves as a user interface, as well as a processor for many time-consuming tasks, allowing the server to devote itself to central storage and other tasks.

Client/Server Database—A variation of the relational model with the tables stored on the server and SQL or some other programming interface residing on the clients.

CLP (Superfund's Contract Laboratory Program)—The program that issues contracts to laboratories for Superfund cleanup of the environment.

Codabar—A numeric only bar code consisting of seven black and white bars. Two bars are wide (code 27).

Code 16K—Multirow (stacked) code in which each symbol can have between 2 and 16 rows or stacks. Each row is separated by a 1 module separator bar, and consists of 18 bars and 17 spaces. The code is analogous to sentences in a paragraph.

Code 39—A full alphanumeric bar code consisting of nine black and white bars. Three bars are wide (code 3 of 9). Code 39 is the most frequently used symbology in industrial bar code systems today.

Code 128—A full alphanumeric bar code capable of encoding all 128 ASCII characters.

Coercivity—Value of the opposing magnetic intensity that must be applied to a material to remove the residual magnetism when it has been magnetized to saturation.

Compatibility—The ability of two devices to communicate with each other understandably; the ability of software to run on a particular hardware platform.

Concatenate—To link together.

Control chart (process control chart)—The graphical representation of process data, consisting of the lines for the control limits that provide a statistical boundary within which the process measurement is expected to fall; and a plot of the measurements of a process characteristic or statistic over time or sequence that can be used to control and improve a process.

Control chart (quality control chart)—A graphical representation of sampling data, consisting of the lines for the control limits that provide a statistical boundary for which the sample measurements is expected to fall within; and a plot of the measurements of a sample characteristic or statistic over time or sequence to improve the control of the quality of a measurement.

Control limits—The upper and lower boundaries that signify a process or a test procedure is out of control.

Control samples—Pooled or composite test samples that are analyzed to check the proficiency of the lab measurements. Control samples in a manufacturing laboratory may also be taken to check a production run.

Control substance—Any material minus the test substance that is monitored in a study for establishing a basis for comparison with the test substance. Controls in a study represent the absence of the test material and are analogous to a blank in an analytical method.

CPS—(Characters Per Second).

CPU (central processing unit)—Controls the operation of the entire computer system and executes the arithmetical and logical functions contained in a particular program.

CRT (Cathode Ray Tube)—Also called a terminal or monitor.

CSA (Computer Security Act)—A list of requirements for computer security.

CuSum chart—The cumulative sum control chart used to detect abrupt changes in the process.

Data—A general term used to denote any or all facts, numbers, letters, and symbols that refer to or describe an object, idea, condition, situation, or other factors. It denotes basic elements of information that can be processed or produced by a computer.

Data acquisition system—Any instrument or computer that acquires data from sensors via amplifiers, multiplexers, and any necessary analog to digital converters. Typically associated with process industries.

Data collection—The act of bringing data from remote points to a central location and organizing it into understandable information. Typically associated with discrete manufacturing.

Data encapsulation—The separation of internal object information from external demands upon the object.

Data matrix—Variable-size two-dimensional matrix symbology that is inherently omnidirectional. Data matrix has a high degree of error correction capability and is used primarily for part marking and tracking.

Data model—A standardized representation of data objects used as a container for transactions, a framework for analysis, and a vocabulary for management.

Data warehouse—A database for query and analysis, as opposed to a database for processing transactions. Separating the two functions improves flexibility and performance.

Database—A collection of structured data that is independent of application.

Database integrity—Validity, consistency, and accuracy of the database.

DBMS (database management systems)—Systems that access data stored in a database and present multiple data views to end users and application programmers.

DCE—(data communications equipment)

Decoder—A software or hardware means of translating bar codes into alphanumeric data. A bar code reader is required to scan the information into the decoder.

Default value—The option selected by a computer in the event of the omission of a definite instruction or action.

Density—Compactness of a bar code that measures the narrowest element (usually in mils) of that bar code.

Depth of field—The distance between the maximum and minimum plane in which a bar code reader is capable of reading symbols.

Design—The stage that specifies the automated and manual functions and procedures, the computer programs, and data storage techniques that meet the requirements identified and the security and control techniques that ensure integrity of the system.

Detection limit—The minimum quantity of a substance that can be measured with a predetermined confidence level for presence of the substance.

Direct thermal—A process in which a set of pins on a print head are selectively heated onto heat-sensitive paper (or label stock). In turn, the paper turns dark and a bar code is formed. Over time, a direct thermal image will eventually fade. Also known as thermal printing.

Documentation—The process of gathering written or electronic information describing, defining, specifying, reporting, or certifying activities, requirements, procedures, or results.

DPI—(dots per inch).

DTE—(data terminal equipment).

EAN—A voluntary, nonprofit standards development association active in numbering, bar coding, and Electronic Data Interchange (EDI) messages for products, services, utilities, and transport units and locations. The EAN system is fully compatible with the universal product code (UPC).

EPA—(US Environmental Protection Agency)

Electronic mail (e-mail) or Internet mail: A method of transmitting text messages and files digitally over communication links, allows users to exchange mail with people all over the world via a unique address.

Entity—Something about which information is stored, either tangible or not, such as an employee or a part on the one hand; an event, an account, or an abstract concept on the other.

Entity Class—An abstract group of entities that share a common description.

Entity Set—A set of entities from a given entity class that are currently in the database.

Ethernet—The standard for local communications networks developed jointly by Digital Equipment Corp., Xerox, and Intel. Ethernet baseband coaxial cable transmits data at speeds up to 10 megabits per second. Ethernet is used as the underlying transport vehicle by several upper-level protocols, including TCP/IP.

Expert system—A computer program that uses knowledge and reasoning techniques to solve problems that normally require the abilities of human experts; software that applies humanlike reasoning involving rules and heuristics to solve a problem.

Facility—The premises and operational unit(s) necessary for operating a LIMS.

Fault tolerance—The ability of a system to execute tasks regardless of strategic components' failure.

FCC—(Federal Communication Commission).

FDA—(US Food and Drug Administration).

Feedback—The return of part of the output of a machine, process, or system to the computer as input for another phase, especially for self-correction or control purposes.

FFO (fixed focus optics)—Utilizes small-aperture and nonmoving lenses to digitize an image over a wider range of distance than the traditional CCD reader.

Fiber optics—A data transmission medium that uses light conducted through glass or plastic fibers. Fiber-optic cables have cores capable of conducting modulated light signals by total internal reflection. Benefits include small diameters, high potential bandwidth, and lower cost than copper.

FIFRA—(Federal Insecticide, Fungicide and Rodenticide Act).

File—A general term for a named set of data items stored in machine-readable form.

Firewall—Typically a separate computer that sits between the Internet and the internal network dedicated to examining traffic for trouble and enforcing strict security measures.

Firmware—Computer programs, instructions, or functions implemented in user-modifiable hardware. Such programs or instructions, stored permanently in programmable read-only memories, constitute a fundamental part of system hardware.

First read rate—The ratio of the number of successful reads to the number of attempts.

Floating-point data—A mathematical notation in which the decimal point can be manipulated. Values are sign, magnitude, and exponent.

Flow chart—Diagrammatic representation of the operations involved in an algorithm or automated system. Flow lines indicate the sequence of operations or the flow of data, and special standard symbols are used to represent particular operations.

Font—A specific size and type of printable character.

Footprint—The amount of floor or table space taken up by a unit or object.

Frequency—The number of recurrences of a periodic phenomenon in a unit of time, usually electronically specified in Hertz (Hz): one cycle per second equals 1 Hz.

Front-end—Operator interface or application-specific aspects of a program.

FTP (file transfer protocol)—A tool available for years to send and retrieve files from other computers on the internet. FTP provides a command line interface that is not elegant, but is straightforward to use.

Function—1. The characteristic actions, operations, or kind of work a person or thing is supposed to perform (e.g., the engineering function or the material-handling function). 2. The operation called for in a computer software instruction.

Fuzzy logic—A method used to model linguistic expressions that have nonbinary truth values. It has been used with PID (Packet Identifier Field) algorithms in process control, especially where process relationships are nonlinear.

GALP (Good Automated Laboratory Practice)—Standards that provide guidance in the use of automated equipment and instruments in the laboratory. The EPA is active in this area.

Gateway—A computing system or software function that performs a protocol or API translation, and serves as an intermediary between computing systems or communication networks.

GLP (Good Laboratory Practice)—Standards that provide guidance in good laboratory techniques and operations that may affect the quality of data measurements. The FDA and the EPA have issued GLPs that are required by companies in their respective areas of responsibility.

GMP (Good Manufacturing Practice)—Standards that provide guidance for producing quality product.

GMP (Good Measurement Practice)—Standards that provide guidance for acquiring quality measurements.

Grade—A ranking of the quality of a product or service based on characteristics or features that are intended for different needs of the consumers of the product or services.

Groupware—A type of software designed to raise the productivity of people working in groups.

GUI (graphical user interface—Software application manipulation by means of windows, icons, menus, and other graphical representations.

Hardware—The physical, manufactured components of a computer system, such as the circuit boards, CRT (Cathode Ray Tube), keyboard, and chassis.

Hierarchical—An approach used in numerous technologies, including machine vision, process control, networking, databases, and planning, whereby the scope of work is arranged in hierarchies that establish priorities and appropriate routings. A database architecture in which data elements are arranged in the form of an inverted tree structure in which no data element has more than one parent.

Host computer—The primary computer in a multicomputer network that issues commands, has access to the most important data, and is the most versatile processing element in the system.

Hot key—A key (or set of keys) designated to activate a specific function in an application program.

HTML (Hypertext Markup Language)—Plain text that contains some formatting instructions (in the form of tags or HTML markup codes) that advise Web browsers how to display and print the documents.

HTTP (Hypertext Transport Protocol)—The protocol of the World Wide Web.

Hypertext—An interactive on-line documentation technique that allows users to select (typically via a mouse click) certain words or phrases to immediately link to information related to the selected item.

IAN (Industrial Article Numbering)—See EAN.

Import/export capability—The ability to read/write data to/from a data file in a known data format for the purpose of exchanging data within software applications.

Information—Any communication or reception of knowledge such as facts, data, or opinions, including numerical, graphic, or narrative forms, whether oral or maintained in any medium, including computerized databases (floppy disk and hard disk), papers, microform (microfiche or microfilm), or magnetic tape.

Infrared laser diode—An invisible light beam used in some bar code readers to scan a bar code that is invisible to the human eye. This technology is used specifically to prohibit people from reading the bar code visually. Analogous to night vision goggles.

Initiation—A request for the development of a system to meet a need for information or to solve a problem for the individual making the request.

Input signal—A signal applied to a device, element, or system.

Inspect—To measure, examine, test or gauge one or more characteristics of an entity and compare the results with specified requirements to establish whether conformance is achieved for each characteristic.

Installation and operation—Incorporation and continuing use of the new system by the organization.

Installation qualification—The act of establishing confidence that a system is capable of operating within stated limits and tolerances.

Instrument—1. Any item of electrical or electronic equipment designed to carry out a specific function or set of functions. 2. device for measuring the value of an observable attribute. The instrument may also control the value.

Instrumentation—Devices, interconnections, and systems used to observe, measure, and/or provide data as to what is occurring or has occurred in order to evaluate or control physical phenomenon or processes.

Integrity—Sound, unimpaired, or perfect condition. The computer security characteristic that ensures that computer resources operate correctly and that the data in the databases is correct. It protects against deliberate or inadvertent unauthorized manipulation of the system and ensures and maintains the security of entities of a computer system under all conditions. Integrity is concerned with protecting information from corruption.

Interface—1. A shared boundary between two pieces of equipment. 2. The hardware and software needed to enable one device to communicate with another.

Interleaved bar code—A bar code in which characters are paired using bars to represent the first character and spaces to represent the second character.

Interleaved 2 of 5—A numeric only bar code consisting of five bars. Two bars are wide, three are narrow. Used generally in industrial and master carton labeling.

Intranet—A company-wide internal system that allows employees and authorized users to access data, information, and bulletin boards via a browser on their computer. There are several advantages to the intranet: it allows for greater communication within the company, from bulletin boards to company-wide mail, and rapid access to historical data for statistical analysis. Of course an intranet has security, with multiple levels of permissions, and can track individual usage.

Internet—Computer network created around 25 years ago by the US Department of Defense (Defense Advanced Research Projects Administration [DARPA]) to allow computers located around the world to transfer information and data robustly and reliably primarily for military purposes. The Internet is a network of networks. Local networks around the world are linked by wires, telephone lines, fiber optic cables, microwave transmissions, and satellites.

ISDN (integrated services digital network)—Method of utilizing existing telephone wire digitally rather than in an analog format. ISDN lines can "move" data at speeds ranging from 64K to 1.544 Mbps. This allows significantly faster internet access than modems.

ISO—(International Organization for Standardization)

ISO 9000 Registration—A certification process in which a registered company will gain international quality assurance recognition for its products.

ISO 9000 standard series—A set of international quality assurance standards.

Key—A key with a property such that if you remove an attribute, the resulting set is no longer a superkey.

Knowledge-based system—Software that uses artificial intelligence techniques and a base of information about a specialized activity to control systems or operations.

Laboratory management—Individuals directly responsible and accountable for planning, implementing, and assessing work, and for the overall operation of a facility.

LAN (Local Area Networks)—Network that spans a limited geographical area to connect computers and terminals, usually at moderate to high data rates.

Laser scanner—An optical bar code reading device using a low-energy laser light beam for scanning.

LASF (Laboratory Automation Standards Foundation)—A foundation established to standardize the best practices of laboratory automation.

LCD—(Liquid Crystal Display)

LED (Light Emitting Diodes)—Solid-state devices that radiate in the visible region used in alphanumeric displays and as indicator or wiring lights.

LIMS (Laboratory Information Management System)—Automated systems used in various industries to improve sample tracking and communications between quality control and production functions.

LIMS raw data (LRD)—Original observations recorded by the LIMS that are needed to verify, calculate, or derive data that is or may be reported.

LIMS raw data (LRD) storage media—The media to which LIMS raw data are first recorded.

Linear—Describing any device or motion in which the effect is exactly proportional to the cause.

Logout—The act of readying a sample for reporting. After all testing of the sample has been completed, any product grading or final sample validation checks are done at logout and the sample is set to a data completion status.

Lot—A quantity of product considered uniform for sampling.

LPM—(Lines Per Minute).

Machine language—Binary instructions to a computer that it can execute directly, without translation.

Macro—A kind of computer shorthand that reduces many commands to one, making it easy to activate frequently used functions.

Mainframe—A large computer normally requiring a controlled environment in terms of temperature, air quality, and static electricity.

Maintainability—The probability that a system can be maintained or restored within a given period of time.

Maintenance/enhancement—Resolving problems not detected during testing, improving the performance of the product, and modifying the system to meet changing requirements.

MaxiCode—Fixed-size, two-dimensional symbology having elements arranged around a unique circular finder pattern. MaxiCode is omnidirectional and is primarily used for freight sorting and tracking.

Memory—A device into which data can be entered, in which it can be held, and from which it can be retrieved at a later time.

MICR (magnetic ink character recognition)—Style of printing on the bottom of personal and bank checks.

Microprocessor—A programmable, large-scale integrated circuit containing all the elements required to process binary-encoded data, and having all the functions of a computer, except memory and I/O (Input/Output) systems. A microprocessor can perform basic arithmetic, and logical as well as control functions equivalent to the CPU of a conventional computer.

Mil 1/1000th—0.001 of an inch or approximately 0.0254 mm. Bar code densities are commonly referred to as number of mils (e.g., 10 mils).

Minicomputer—A class of computer having a CPU constructed of a number of discrete components and integrated circuits, rather than being comprised of a single integrated circuit, as in a microprocessor. A "mini" is larger than a microcomputer and has a typical word length of 16 or 32 bits. It is a small, programmable, general-purpose computer typically used for dedicated applications.

Misread—When the data output of a bar code reader does not correspond with the data encoded in the bar code symbol.

Modeling—The mathematical characterization of a process, object, or concept to enable the manipulation of variables so as to simulate typical behavior in programmed situations.

Modem—A MODulator–DEModulator that interfaces with data processing devices to convert data to a form compatible for sending and receiving over transmission facilities, most commonly telephone lines.

Module—The narrowest nominal unit of measure in a bar code.

Motherboard—A printed circuit board that holds the principal components of a microcomputer.

MRD (minimum reflectance differential)—A method that is used to determine if there is an adequate difference between absorbed and reflected light.

MRP (material requirements planning)—A system that typically tracks materials, keeps inventory, generates bills of material, work orders, and scheduling.

MSI Plessey (modified Plessey code)—A pulse-width-modulated bar code used primarily for marking retail shelving.

Multiplexer—An I/O device that routes data from several sources to a common destination.

Multiprocessing—1. Pertaining to simultaneous execution of two or more programs or sequences of instructions by a computer. 2. Use of a linked set of central processors to perform parallel processing.

Multitasking—Procedures in which several separate but interrelated tasks operate under a single program identity.

Natural language—Any naturally evolved human language.

NBS—(US National Bureau of Standards), renamed NIST (National Institute of Standards and Technology).

Nesting—1. To embed a subroutine or data block in a larger routine or data block. 2. The organization of data in hierarchical structures for greater efficiency in storing and processing repetitive elements. Identical elements therefore need to be represented only once in a database.

Network—1. Any system of computers and peripherals. 2. In an electrical or hydraulic circuit, any combination of circuit elements.

Network computer—An extremely "thin" computer that sells for as little as $500 and uses Web access to acquire data and functionality.

Network management—In a client/server architecture, making sure all hardware, servers, PCs, hubs, switches, bridges, routers, and other equipment are working properly.

Network topology—The physical arrangement of communication nodes in a network.

Neural network—A processing architecture derived from models of neuron interconnections of the brain. Unlike typical computers, neural networks are supposed to incorporate learning, rather than programming, and parallel, rather than sequential, processing.

NFS (Network File System)—A protocol that allows a computer to utilize disk space and files of another computer over a TCP/IP network.

Node—One component of a network where interconnections occur.

Noise—In general, any unwanted disturbance superimposed on a useful communications signal that tends to obscure information content.

Nominal—The ideal value for a specified parameter.

Null modem—A cable that criss-crosses the DTE and the DCE signals between two DTE devices, so that the two devices can communicate.

Object-oriented database—A database used to store objects that form the basis of object-oriented computing, in which data, as well as references to the procedures used to perform operations on that data, are combined.

Object-oriented programming—Programming based on objects that communicate by passing messages. An object is a package of information and descriptions of procedures used to manipulate that information.

Object-oriented software—Results from a kind of modular programming. Each object is a software package containing a collection of related procedures and data that can be reused to shorten application development time. Objects also make it easier to customize software systems to mirror actual business processes without having a negative impact on the ability to migrate to later software releases.

OCR (Optical Character Recognition)—A technology designed specifically to read certain stylized fonts (such as OCR-A and OCR-B) containing the full alphanumeric character set. The term OCR is also used when translating and inputting other stylized fonts (such as Courier) or text found in magazines and newspapers into a computer. This technology is also referred to as intelligent character recognition (ICR).

OCR-A—Character set used for optical character recognition and described in ANSI Standard X3.17-1981.

OCR-B—Character set used for optical character recognition and described in ANSI Standard X3.49-1975.

Off-Line—1. Any element of a process that stands independent of its normal flow. 2. Describing the state of a subsystem or computer that is operable, but currently bypassed or disconnected from the main system. 3. Devices not under direct control of the CPU.

OIRM (Office of Information Resources Management)—An office within EPA that has responsibility for GALPs.

On-Line—1. Process elements that are an integral part of its normal flow. 2. A subsystem or computer that is operable and connected to the main system. 3. Devices under direct control of the CPU.

Open systems—An approach to computing that stresses the interconnectability of systems based on compliance to established standards.

Operating system—A structured set of system programs that controls the activities of a computer system and associated peripheral devices, as well as the execution of programs and flow of data.

Operational qualifications—The act of establishing confidence that a system in operation is performing satisfactorily and within specifications.

Operator interface—The shared boundary between the human operator and a computer system, typically consisting of a graphical representation, keyboard, or mouse.

Output—1. The end result of a process or system. 2. Information leaving a device; data resulting from the processing. 3. An audio, electric, or mechanical signal delivered by an instrument to a load.

Parallel processing—Processing performed by a computer with two or more CPUs that execute small sections of a task in parallel, resulting in improved performance.

Parameter—A variable in terms of which it is convenient to express other interrelated variables that may then be regarded as dependent.

PC (personal computer)—An inexpensive computer system that serves the needs of a single user, typically for business or productivity applications.

PCS (Print Contrast Signal)—Measurement of the ratio of the reflectivities between the bars and the spaces of a symbol.

PDF417—Two-dimensional bar code providing error correction, detection, and security, used primarily in parcel tracking applications and hazardous material control.

PDT (Portable Data Terminal)—Hand-held terminal capable of storing and recording data that is captured remotely and later transmitted into a computer.

Persistence—The ability of an object to remain in existence past the lifetime of the program that creates it.

Pick list—Also called pop-up. A list of choices from which the user selects.

Pitch—Rotation of a bar code symbol about an axis parallel to the direction of the bars.

Pixel—The smallest element of controllable color and brightness in a video display or computer graphic.

Plotter—An output device that converts computer output into lines drawn on paper or on display terminals.

Polymorphism—The ability of a class to respond to a common message or operation in a distinctive manner.

Population—A set of data that represents all possible values of a characteristic or event under consideration.

Port—The point at which signals from peripheral equipment enter a computer.

PostNet (postal numeric encoding technique)—Used to encode ZIP code information on letter mail. PostNet utilizes redundant information within a compact bar code format to provide error detection capability and a significant degree of error correction capability.

Precision—A measure of the degree of accuracy between independent test results of the same sample.

Printhead—The device on a direct thermal or thermal transfer printer containing the heating element that causes an image to be transferred to the face stock.

Process—A natural phenomenon marked by gradual changes that lead toward a particular result. A series of actions or operations leading to an end. A continuous operation or treatment in manufacturing. Continuous and regular production executed in a definite, uninterrupted manner.

Process capability—Accounting for product specifications and process control.

Process capability index—The ratio of the specification width to the process spread.

Process control—Automatic monitoring and control of a process by an instrument or computer programmed to respond appropriately to feedback from the process.

Productivity—A measurement of output per hours worked.

Program—1. A complete, structured sequence of program statements directing a computer to implement an algorithm. 2. All software programming necessary to solve a problem.

Programming—Coding of the program modules that implement the design.

Programming language—An artificial language that enables people to instruct machines. Computer commands that form procedures by which software programmers design and implement computer software programs.

Protected field—A viewable data field that cannot be changed by the end-user.

Protocol—A standard set of procedures that allows data to be transferred among systems.

PSP (protocols for specific purposes)—Used to define the quality assurance plan of a project or a monitoring program.

Quality—Demonstration of the characteristics and attributes that meet the specifications of a product or service.

Quality assessment—All the activities that monitor and evaluate the performance of quality control procedures used in the production of products or services.

Quality Assurance—All planned activities necessary to provide a high degree of confidence in the quality of a product or service. It provides quality assessment of the quality control activities and determines the validity of the procedures for determining quality.

Quality assurance unit—Any person or organizational element designated by laboratory management to monitor the LIMS functions and procedures.

Quality audit—The systematic examination and evaluation of all activities related to the quality of a product or service to determine the suitability and effectiveness of the activities to meet quality goals.

Quality control—The operational techniques and activities needed to maintain and improve the quality of a product or service.

Query—A request for data that is initiated while a computer program is running.

Quiet zone—A clear space, containing no dark marks, preceding the start character of a symbol and following the stop character.

RAM (Random Access Memory)—Computer memory that can be read from and added to by the programmer.

Random error—The error produced in a measurement process due to causes that are indeterminate or nonassignable.

Random sample—A sample obtained in a manner such that all samples of a population have an equal chance of being selected.

Range—The set of values within which measurements can be made without changing the measuring instrument's sensitivity.

Raw data—Data that cannot be calculated or derived from other information.

RCRA (Resource Conservation and Recovery Act)—The list of disposal regulations for laboratories handling hazardous or toxic substances.

Real time—Response to events in a predictable and immediate way. A spreadsheet response in 1 or 5 secs is acceptable, but closed-loop control systems need to know real-time response rates in a more rigorous manner.

Records—All books, papers, maps, photographs, machine-readable materials, or other documentary materials, regardless of physical form or characteristics, made or received by an agency of the United States government under Federal law or in connection with the transaction of public business and preserved or appropriate for preservation by that agency as evidence of the organization, functions, policies, decisions, procedures, operations, or other activities because of the informational value of the data in them.

Referential constraint—The requirement that each value in the foreign key is a value in the referenced key.

Referential integrity—Making sure that there are no dangling references left in the database when information is deleted.

Reflectance—The light reflected back from the white spaces of a bar code during scanning.

Relational database—A finite collection of tables that provides an implementation of an entity–relationship database model.

Reliability—The ability to perform the required functions of the product or services for a specific period of time and under a specific set of conditions.

Remote—Located a considerable distance from the computer or processing instrument.

Repeatability—The closeness of agreement between individual results, using the same method, test substance, and set of laboratory conditions.

Replication—A database feature that enables information on the network to be constantly updated across several separate computers.

Reproducibility—The closeness of agreement between individual results using the same method and test substance, but a different set of laboratory conditions.

Requirements analysis—Determination of what is required to automate the function(s) identified by the organization.

Resolution—See *Density*.

Retirement—The stage that ends use of the system.

Retrospective validation—The validation of a system already in place using historical data, testing, and control data.

RF (radio frequency)—Wireless communication technology using electromagnetic waves to transmit and receive data. RF provides real-time access to a host computer.

Ribbon—Material used with thermal transfer printers producing visible marks on a label (or substrate). A printhead is heated and the ribbon is burned onto the label stock producing the bar code.

Robotics—The study of the design and use of robots, particularly for their use in manufacturing and related processes.

Rollback/recovery—Ensures that in the event of a partially updated record, the integrity of a database is protected by recovering, or rolling back to the record as it existed before the partial update. This prevents the proliferation of "orphans," which may in turn lead to a corrupted database.

ROM (read only memory)—Computer memory that may not be written by the programmer. The software in the ROM is fixed during manufacture.

RS—232, 422, 423, 449 are the standard electrical interfaces for connecting peripheral devices to computers.

RS-232—The most common communication interface (e.g., serial (COM) port) standard using DTE and DCE interface. Also known as serial communication.

RS-422—The second most common communication interface standard, which extends beyond the 100 feet limitation of RS-232 data communication supporting a maximum distance of 4000 feet.

Ruggedness—A measure of the lack of susceptibility of a test method to changes in environmental factors.

Rules-based system—A functional system in which knowledge is stored in the form of simple if/then or condition/action rules.

Sampling—1. Measuring the output or variable of a process at regular intervals to estimate characteristics of the process. 2. The conversion of a continuous image into an image composed of discrete parts.

SCADA (supervisory control and data acquisition)—Typically accomplished in industrial settings by means of PCs using special software designed for the task.

Scan—To examine data from process sensors for use in calculations. A single sweep of PLC applications program operation. The scan operates the program logic based on I/O status, and then updates the status of inputs and outputs.

Scan time—The time required to execute a Programmable Logic Controller (PLC) program once completely, including an I/O update.

Scanner—A device that electro-optically converts bar codes into electrical signals.

Security—The set of laws, rules, and practices that regulate how an organization manages, protects, and distributes sensitive data.

Semiconductor—Any class of solids having higher resistance than a conductor, but lower resistivity than an insulator. These are the basis for all integrated circuits.

Serial—Relating to or being a connection in a computer system in which the bits in a byte are transmitted sequentially over a single wire.

Serial communication—See *RS-232*.

Server—A processor that provides a specific service to the network. In a client/server architecture, servers perform central storage and other vital tasks. Servers specialize in high-speed computation and data storage and manipulation. They can range in size from PCs to mainframes.

Software—The entire set of programs, procedures, and related documentation associated with a computer.

Software life cycle—The period of time beginning when a software product is conceived and ending when the product no longer performs the function for which it was designed.

Software version control—Management of changes or revisions to a specific baseline software module or application. Software version control provides a mechanism to control changes and to return to any previous revision of the application or module.

SOP (Standard Operating Procedure)—A written procedure for operations performed repeatedly.

Source Code—A software program written using a programming language. It must be assembled, compiled, or interpreted before it can be executed.

SPC (Statistical Process Control)—The use of statistics to control and improve a process.

Specifications—Written requirements for a product, system or service.

SQC (Statistical Quality Control)—The use of statistics to improve and control quality.

SQL (Structured Query Language)—Query language that allows a client to access only the data required to satisfy a request for data, cut network traffic, or improve performance. An accepted standard that most relational database products implement.

Stability testing—A series of product tests conducted at specific time intervals and varying the environmental conditions to see if the product degrades over time.

Stack code—Two-dimensional bar code in which linear bar codes are stacked one upon another and are printed in a rectangular shape to achieve the most efficient use of label area.

Standard—A substance with a known value for a property that can be used to evaluate a property in another substance.

Standard deviation—The positive square root of the variance of a data set; a measure of the spread of a sampling statistic.

Standardization—The promotion of conformity by means of a standard or to establish criteria for uniform practices.

Start–stop character—The left-most and right-most characters of a horizontal bar code that provide the scanner with start and stop reading instructions as well as scanning direction.

Statistic—A value for a parameter calculated from sample data.

Substrate—The surface on which a bar code is printed.

Superkey—A set of attributes for an entity class that serves to identify uniquely an entity from among all possible entities in that entity class.

Symbol—A set of characters and markers including start/stop, quiet zones, data, and check characters required by a particular symbology that form an integrated readable (scannable) element.

Systematic error—A consistent error of the same size and sign produced in a measurement process due to the same recurring cause.

Systems integration—The ability of computers, instrumentation, and equipment to share data or applications with other components in the same or other functional areas.

Table—An array of attribute values whose columns hold the attribute and the rows hold the attribute values for the given entity.

Table Scheme—All of the attribute names for an entity class, for example, store number, product name, cost.

TCP/IP (Transmission Control Protocol/Internet Protocol)—The Internet's communication standard allows different types of computers to share data on a network. IP defines the routing between computers connected to the internet. TCP defines how data is packaged to be delivered by IP. Every transmission gets broken down into standard-sized packets, like little envelopes of data. Each packet carries an address but no information about what is inside.

Terminal—Any I/O device used to communicate with a computer from a remote location.

Terminal emulation—In data collection, using modified business system software screens on data collection terminals.

Testing—The examination of the behavior of a program by executing the program on sample data sets.

Testing and quality assurance—Ensuring that the system works as intended and that it meets applicable organization standards of performance, reliability, integrity and security.

Thermal—See *direct thermal*

Thermal transfer—A process by which a set of pins on a print head are selectively heated onto a ribbon and the ink from the ribbon is burned (transferred) onto the label stock. Thermal transfer leaves a permanent image on the label.

Throughput—1. The rate at which work proceeds through a manufacturing process. 2. The rate at which information is processed through a computer.

Timestamp—A record of the date and time of data entry.

Token ring—The token access procedure used on a network with a sequential or ring topology.

Tolerance—The allowable deviation of the value of a characteristic of a population.

Total quality—A holistic approach to quality control that stresses building manufacturing processes that force users to confront quality problems, rather than passing them on.

Touch screen—An enhanced CRT screen with which an operator can interact by touching icons displayed on the screen, rather than through a keyboard or mouse.

Traceability—The ability to trace the history, application, or location of a substance throughout its life cycle for the purpose of establishing accuracy of measurements.

Trackball—An input device with a ball recessed in its surface, which rotates to control the position of the cursor.

Transaction—1. A computerized record of a discrete event, such as the receipt of inventory or a customer order. 2. A set of two or more database updates that must be completed in an all or nothing fashion.

Transaction logging—Method that provides recovery protection if a failure occurs as data are actually written to the database during the transaction.

Transaction processing—Grouping related elements together, requesting a group lock, and writing any changes to the database as a unit at the end of the transaction.

Triggers—User-defined conditions that automatically initiate specific user-defined responses. For example, inventory dropping below a specific level can be defined as the trigger for automatically generating a purchase requisition.

TSCA—(Toxic Substance Control Act).

Turnkey system—Equipment or a computer system that is delivered complete, installed, and ready to be used.

Two-dimensional bar code—Two-dimensional symbology composed of rows of data arranged in a rectangular or square pattern. The rows of data are stacked onto each other to encode an array of data.

Type I error (alpha error)—Rejection of a hypothesis when it is true. An example of type 1 error is finding a substance present when it is not.

Type II error (beta error)—Acceptance of a hypothesis when it is false. An example of type II error is not finding a substance present when it is present.

UCC (Uniform Code Council)—an organization that administers the UPC and retains other standards.

UPC (Universal Product Code)—Standard bar code symbol for retail packaging in the United States.

Uncertainty—The range of values bounded by credible limits wherein the true value is estimated to lie. It represents the best estimate of the inaccuracy and imprecision of a test method due to random and systematic errors.

UNIX—An operating system developed at AT&T Bell Laboratories and written in C language, which is especially well suited to networking applications.

UPS (Uninterruptible Power Supply)—Used to ameliorate the effects of poor electrical power quality, including voltage anomalies, high frequency noise, or ground loops. It is especially applicable where power outages of more than half-second duration occurs.

URL (Uniform Resource Locator)—A pointer that points to a specific bit of information on the Internet. Each web site has a unique address, called the URL, such as http://www.LIMS.org

User prompt—A message that appears on screen to guide the user to the next available option.

Validation—The process of ensuring that a sample, measurement method, production process, a computer system (LIMS) or data result will meet specifications and conform to predefined quality assurance criteria.

Validity—A state or quality of software that provides confirmation that the particular requirements for a specific intended use are fulfilled.

Variance—A measure of the spread of data. It is the averaged squared deviation about the mean.

Verifier—A device that measures the characteristics of a bar code including the contrast, reflectance, modulation, and compliance with the parameters of the bar code symbology.

Verify—To review, inspect, test, check, audit, or otherwise establish and document whether LIMS raw data are accurate.

VGA (video graphics array)—Introduced by IBM to supersede EGA by offering 640 × 480 pixel resolution.

Visible laser diode—Used in most hand-held scanners to project a visible red light for scanning human-readable bar codes.

Wand—A penlike scanning device used as a contact bar code reader.

Web browser—A user-friendly graphical user interface that can be thought of as a window to the World Wide Web. Web browsers now support integrated or linked graphics, video, and audio clips. Browsers also allow users to send e-mail, access FTP sites to download software or data files, and access newsgroups.

Wedge—A hardware device or software program (software wedge) that uses a scanner for input and sends data directly into an application by emulating a keyboard stroke. A hardware wedge is an external device and plugs between the keyboard and terminal.

Windows—1. A temporary, bounded area on a computer screen that is user specified to include data particular to a given application. 2. An MS DOS-based graphical user interface.

Workflow—The ability to graphically designate and change the distribution and approval routings of documents related to business process.

Workstation—1. An area used in a manufacturing process to perform a series of functional tasks, usually associated with a single operator. 2. A single-user computer, typically with 32-bit messaging and integrated graphics.

World Wide Web—A global, interactive, distributed, cross-platform, graphical hypertext information system that operates over the Internet. An Internet system for hypertext, and the linking of multimedia documents, allowing users to move from one Internet site to another and to inspect information that is available without using complicated commands and protocols.

Worst case—The set of conditions using the upper and lower boundary limits for checking the functionality and operations.

WYSIWYG (what you see is what you get)—The presentation of printable output viewed as is on the terminal, pioneered by Apple Computer.

X dimension—Dimension of the narrowest bar in a bar code.

XML—A markup language for documents containing structured information. Examples of structured information include both content (words, pictures, etc.) and some indication of what role that content plays (for example, content in a section heading has a different meaning from content in a footnote, which means something different than content in a figure caption or content in a database table, etc.). Almost all documents have some structure. A markup language provides a means to identify structures in a document.

LIMS Suppliers

LIMS Vendor	Product	Underlying Database	Web Interface Supported	Web Address
Accelerated Technology Laboratories, Inc.	Sample Master Pro ScreenIT LIMS	MS SQL Server, ORACLE, Access 2000	Yes	www.atlab.com
Applied Biosystems	SQL LIMS	Oracle	Yes	www.appliedbiosystems.com
Autoscribe Ltd.	Matrix Plus	Oracle, MS SQL Server, DB2	Yes	www.autoscribe.co.uk
Baytek International, Inc.	WinBLISS	Oracle	Yes	www.baytekinternational.com
Beckman Coulter, Inc.	LabManager iLIMS	Oracle	Yes	www.beckmancoulter.com
Blaze Systems Corporation	BlazeLIMS	Oracle, MS SQL Server	Yes	www.blazesystems.com
ChemWare	Horizon LIMS	Oracle	Yes	www.chemware.com
Cogexel, Inc.	LabPlus	Oracle, MS SQL	Yes	www.cogexel.com
Creon Lab Control, Inc.	Q-DIS QM-AQV-DM-R	Oracle, MS SQL	Yes	www.creonlabcontrol.com
LabVantage Solutions	LV LIMS Sapphire Edition	Oracle, MS SQL Server	Yes	www.labvantage.com
LabWare Inc.	LabWare LIMS	Oracle, SQL Server, Sybase, Informix, DB2	Yes	www.labware.com
L.I.M.S. (USA), Inc.	StarLIMS	Oracle, MS SQL, SQL-based DBs, Sybase	Yes	www.starlims.com
QSI Corp	WinLIMS & WebLIMS	Oracle, MS SQL, SQL-based, Sybase	Yes	www.qsius.com
RJ Lee Group	Lab Task	Btrieve/Pervasive SQL	Yes	www.rjls.com
Thermo LabSystems, Inc.	Sample Manager, Nautilus	Oracle	Yes	www.thermolabsystems.com
Zumatrix Inc.	Matrix Plus	Oracle, MS SQL Server, DB2	Yes	www.zumatrix.com

Appendix A
Sample Request for Proposal

I. INFORMATION FOR PROPOSERS

1.01 SCOPE

A.   Company A, Water Quality Department, is seeking proposals from qualified vendors to provide a laboratory information management system
(LIMS).

     The company is seeking Proposers who can provide a complete
solution that meets their current and future LIMS software requirements.
The LIMS will encompass one main laboratory. The company requires
a mature client/server solution with a proven graphical user interface.

B.   There are three parts of this request for proposal (RFP): Information for
Proposers, Proposal Preparation Instructions, and the Scope of Work.

1.   The Information for Proposers briefly describes the scope, defines
terms, and provides specific information about submission of the
proposal, the evaluation criteria, contract award, payment, and other
requirements.

2.   The Proposal Preparation Instructions describes the proposal format
and what information is to be included in each proposal section. A
Price Schedule format is provided.

3.   The Scope of Work contains the Technical Specifications for the
LIMS, including installation, configuration, support and training services, and acceptance criteria; a Table of Compliance to Specifications; and a Questionnaire.

1.02   COPIES OF PROPOSAL DOCUMENTS

A.   The following three sections of this document are provided on the en-
closed diskette in Microsoft Word 6.0. These are provided for the conve-
nience of the Proposer, to make it easier to complete and return printed
versions of these sections using the required format.

Proposal Preparation Instructions: Price Schedule
Scope of Work: Table of Compliance
                                                      Questionnaire

B.   Direct all technical questions about the meaning or intent of the Request
for Proposal to _____.

C.   The proposal shall follow the format specified in the Proposal Preparation Instructions Section and shall be delivered in a sealed envelope by
close of business on _____. No oral or electronically transmitted proposals will be considered. The proposal and each of its schedules shall be carefully filled out and signed by the Proposer. The sealed
proposal shall be clearly marked with the proposal name shown on the
request for proposals and shall be delivered at the time and locations
specified in the request for proposals. Proposals received late will not
be considered.

1.03   PROPOSAL PRICE SCHEDULE

A.   Submit prices as defined in the Proposal Preparation Instructions, Price
Schedule, which is included in this document. Unit prices shall be shown
for each unit specified and shall include all shipping charges, and all
tariffs and excise taxes.

B.   Prices must be submitted for all items in the Bid and the additive Option.
Company A may elect to accept any combination, all, or none of the
Options. The Base Bid shall be based on the minimum requirements and
specifications contained in this RFP. Prices for any proposed/recom-
mended enhancements to the minimum should be included in the Additive Options.

C.   Company A reserves the right to exclude certain items and services included in the Base Bid after the proposal date. This shall be considered
in the listed prices and all proposals will be evaluated equivalently.

1.04 SUBMISSION OF PROPOSAL

A. All proposal packages must be submitted in a sealed envelope.

B. Proposals must be received by no later than close of business on ____.

C. Submit three copies of the proposal to:

D. In addition to printed copies, the Proposer is encouraged to include one
(1) electronic version on diskette.

E. It is the responsibility of all Proposers to examine the entire request for
proposal package and seek clarification of any requirement that may not
be clear and to check all responses for accuracy before submitting a
proposal.

F. Only complete responses will be considered. Proposals that fail to address all software and services required by the Technical Specifications
will be judged nonresponsive and will not be considered.

1.05 EVALUATION CRITERIA

A. In evaluating Proposals, Company A will consider the following criteria:

1. Unit prices and total cost of Base Bid. Additive Options will also
be considered pending availability of funds and technical evaluation.
All proposals will be evaluated equally with the same combination
of Base Bid and desired Additive Option.

2. The performance, reputation, financial stability, qualifications, and
experience of the Proposers, Suppliers, and other persons and organizations proposed for the Work. Evidence of Proposer's ability to
meet these criteria must be submitted as part of the Proposal, and
will include the names and telephone numbers of references, as well
as evidence of financial stability and business reliability.

3. The technical merit of the Proposal including compliance with the
prescribed requirements, and any enhancements. The extent to which
the Proposer's proposed softwear products exceed the specified requirements also will be used as a basis for evaluation. This may
include configuration suggestions that increase the price/perfor-
mance benefits.

4.   The schedule will be considered as a measurement of commitment and ability to perform according to the needs of Company A.

B.   The criteria outlined above will be weighted as follows:

1.   Cost   25%
2.   Company Performance, Business Reliability, Qualifica-   20%
     tions, and Experience
3.   Technical Merit of Proposal   50%
4.   Schedule   5%

C.   Proposer Short List: The Company A will select a short list of Proposers based on the evaluation described above. The short-listed Proposers will be invited to demonstrate their LIMS.

D.   Product Demonstration

1.   The LIMS products shall be presented using a scripted demonstration provided.
2.   This scripted demonstration will be provided to each short-listed Proposer within three (3) working days following selection.
3.   There will be a minimum of two (2) weeks before the demonstrations.
4.   Selection of the successful Proposer will based on the evalution of the demonstrated LIMS.

1.06   ANTICIPATED SCHEDULE

A.   The Following schedule is presented for information purposes only. All dates and indications of time are tentative and subject to change of a.

Issue request for proposals   _____
Proposals due   _____
Selection of short list   _____
Vendor demonstrations   _____
Selection of successful proposer   _____
Purchase of LIMS   _____
Installation of LIMS/training   _____

B.   Company A reserves the right to reject any and all Proposals and to disregard all nonconforming, nonresponsive, unbalanced, or conditional

Proposals. Also, Company A reserves the right to reject the Proposal of any Proposer if it believes that it would not be in the best interest to make an award to that Proposer, whether because the Proposal is nonresponsive, the Proposer is unqualified or of doubtful financial ability, or the Proposer fails to meet any other pertinent standard or criteria established herein. Proposals will be analyzed using the criteria indicated in this section contained herein to determine the best value.

1.07 PAYMENT TERMS

A. Payment terms shall be negotiated with the successful Proposer.

II. PROPOSAL PREPARATION INSTRUCTIONS

1.01 GENERAL

The Proposer shall submit the proposal in a sealed volume. The following items shall be submitted with the proposal.

1.02 TECHNICAL DESCRIPTION

A. The technical description shall demonstrate a thorough understanding of the requirements of the specification and a logical plan for accomplishing the contract requirements. Elaborate format and binders are not necessary. Legibility, clarity, and completeness of the technical approach are much more important.

B. To aid in the evaluation of the technical description, it is required that all proposals follow the same general format. Therefore, prepare the proposal in accordance with the following format and, as a minimum, include the information specified under the format headings.

1. Table of Contents
2. List of Figures and Tables
3. Technical Approach. Each Proposer shall include a narrative describing the Proposer's recommendations, methods, and techniques

for accomplishing the tasks listed in the Systems Specifications. This narrative shall include a description of any area not addressed in the Systems Specifications that the Proposer believes to be essential to successful completion of the project. If and when the Proposer's methodology differs from the concepts described in this document, the Proposer shall describe the differences. The detail provided is of great importance in aiding with the evaluation of the proposal. Any Proposal failing to address itself clearly and completely to the specifications may be considered nonresponsive.

4. Table of Compliance to Specifications. The Proposer shall complete the included compliance table. The following definitions shall be used when completing the table.

A. "COMPLY." Where the proposal is in compete accordance with the Specification statement.

B. "DO NOT COMPLY." Where the proposal does not meet the specificiation requirements and no alternative is proposed because of a prohibitive development price or schedule delay.

C. "COMPLY WITH MODIFICATIONS." Where the proposal varies from the Specification requirements, the Proposer will use the format described in the instructions for completion of this table to provide explanation of the deviation, including a reference to the Specification paragraphs involved.

Explain how the proposed equivalent meets the functional intent of the Specification and submit documentation describing the substituted item.

Should a point of conflict develop between the proposal and the specifications and formal written exception is not taken before contract award, then the dispute shall be resolved in favor of the specifications. If no exceptions are listed, it is assumed that the Proposer is in complete conformance with all specified requirements and will be required to perform accordingly.

5. Questionnaire. The Proposer shall answer all the questions contained in the Questionnaire following the format provided.

6. Prior Experience and Performance. The Proposer will include the description of at least three projects or installations of a similar nature of work performed in the past or currently on-going, which would substantiate the qualifications of the Proposer for this project.

Company A will contact the Proposer's previous clients to validate the accuracy of all statements of qualification.

The Proposer will include the following for each project/site:

a. Name and description of project
b. Contract amount
c. Name of client
d. Contact person and telephone number
e. The type of business (i.e., water utility lab, wastewater lab, etc.)

7. Schedule. The Proposer shall include a project schedule with the Proposal specifying the duration, in calendar days, for the procurement of the software as defined in the Scope of Work, and the proposed duration of the installation, configuration, and training for the software systems. If any customization is required, include this explicitly in the schedule.

8. Evidence of Qualifications. The Proposer shall include evidence of financial stability, qualifications, and acceptance or exceptions to the terms and conditions of the Contract Documents.

9. Miscellaneous Information. Additional information to clarify or augment proposals for the Company A project is acceptable but brevity is important.

1.03 PRICE SCHEDULE

A. The Proposer shall include a Price Schedule with the Proposal. All prices submitted shall be considered firm over the life of the contract. Late delivery or underestimation of the work required for the system shall not be justification for price adjustment.

B. Costs on the Proposer's Price Schedule are to be indicated and summed by the following categories:

1. Software: The Proposer shall include costs and subsequent license fees for all software products deemed necessary to meet the functional requirements described in the Scope of Work in their response to this RFP. This includes all base system/modules, optional systems/modules and all supporting software, both required and optional. Include both Server and Client licenses. Provide individual line item cost for interfacing each listed instrument.

2. Services: The Proposer shall provide line items for their installation, configuration, customization, and training services and the associated cost of each service with their proposal. All costs to Company A to such as labor, travel, and expenses must be included.

3. Hardware: The Proposer shall specify all extra or special hardware or equipment needed to meet RFP requirements defined in the Scope of Work. This is to ensure that the electronic instrument interfaces function correctly.

III. SCOPE OF WORK

PART 1   LIMS TECHNICAL SPECIFICATIONS

1.0 Overview and Purpose

A.   General Specifications

These specifications identify minimum and optimal functional and processing capabilities required for the computerized LIMS.

1. A computerized LIMS shall provide Company A's water quality laboratory with management information tools to allow for efficient laboratory operations in producing timely and accurate analytical data and assessment reports, and to make validated data available to all required parties. Data entry, access, and retrieval shall be provided, at a minimum, for the following:

   - Manual data input by users
   - Direct data acquisition from laboratory instruments
   - Data storage
   - Data processing and manipulation
   - Data retrieval and reporting

2. The LIMS processing functions shall incllude the following:

   - System management
   - Database management
   - Sample management and tracking
   - Workload management
   - Sample analysis and data acquisition
   - Data validation and limit checking

- Quality control/assurance
- Statistical data analysis and graphics
- Data import/expert capability
- Ad-hoc querying
- Barcoding
- Reporting

3. The LIMS shall perform data acquisition from laboratory instruments, while simultaneously supporting workstations on a Novell network performing other LIMS functions.
4. The LIMS application software shall be comprised of proven packages. These packages shall permit on-site configuration and generation of all application related programs including displays and reports.
5. The LIMS application software shall be a standard product that is fully developed, tested, and supported. It shall be compatible with the system hardware, and shall meet the functional requirements specified.
6. All system software shall be designed to allow growth. Sufficient space shall be recommended to allow for additional screen displays, and for additional, or expanded, reports.

B. System Configuration

1. Network
   The LIMS shall be installed on the Ethernet network, with the Novell 3.12 or 4.x network operating system.
2. Database Server
   Company A will supply and install a database server for the LIMS application. The LIMS shall run on this server, configured in client/server mode. The server will meet the following general specifications:

   - Intel 200 MHz Pentium with 256 MB RAM
   - Windows NT Server
   - A mirrored 4 MB wide SCSI hard drive
   - CD ROM drive

3. Personal Computers
   The client workstations are 133 MHz Pentiums with 64 MB RAM that will be dedicated to the LIMS

2.0 LIMS Requirements

A. System Management

1. Licensed users
   The LIMS shall be licensed for 8 concurrent users not counting the interfaced instruments. Up to 10 workstations shall have access to the LIMS.

2. Compatibility
   The LIMS shall run on a server platform and an operating system compatible with the existing Novell NetWare 4.x.

3. System Management Tools
   The LIMS shall provide system management tools to permit safe, secure management of the LIMS application. These tools shall include application security, data audit trail, database backup/recovery, data archival/restoration and interoperatability with SQL-based and ASCII-based applications.

4. Security
   The LIMS system shall provide security features to ensure that only authorized individuals enter, view and modify data. Access levels shall be definable to restrict use of system level functions (such as user authorization), and to provide data access levels to restrict the use of data entry, data approval, data retrieval, data modification, database structure creation or modification functions.

5. Data Archiving and Purging
   The LIMS shall provide a means to archive and purge (delete) data at the request of the system administrator, or automatically after a specified period of time.

   - Archiving is removing the data from the active database and storing it in a retrievable form elsewhere. Archiving must include user-selectable parameters. These parameters shall include collection and approval date ranges, sample type, location, and test.
   - The purge utility must also include user-selectable parameters. These parameters shall include collection and approval date ranges, sampling point and sample type.

6. Static Information
   The LIMS shall maintain static administrative information such as, but not limited to, procedures, safety information, and project infor-

mation. Authorized users shall be able to query, add, modify, and delete this information.

B. Database Management System

1. Relational Database Management System
   The LIMS shall provide a relational database management system (RDBMS) for information storage and retrieval.

   - The LIMS RDBMS shall be available with a full-use license, providing not only access to the LIMS application, but also application development tools, a data dictionary, a data query utility, and a report writer. The preferred databases are ACCESS or SQL. Oracle systems will not be considered.
   - The RDBMS shall be licensed for eight concurrent run-time users. The database development tools shall be licensed for two users. The report writer tools shall allow development by five concurrent users.
   - The RDBMS shall support client/server architecture.
   - The RDBMS shall support parallel processing.
   - The RDBMS shall be able to support data spanning multiple physical disks.
   - The RDBMS shall run on multiple server operating systems, such as Windows NT or Novell NetWare.

2. Transaction Journal Utility
   A transaction journal utility shall provide database reconstruction in case of system failure. This facility shall restrict the possible loss of data to the database transactions in progress when the system fails. Proposer must provide written instructions for reconstruction.

3. Graphical User Interface
   The LIMS user interface and all interactive database management tools shall be a simple-to-use graphical user interface (GUI).

4. Data Export
   The Database System shall be able to extract and convert data elements into an ASCII format for use outside of the LIMS application environment. The following file formats are desired or required, as indicated:

   ASCII: Required
   EXCEL: Required
   Lotus: Desired

5. Data Import
    The Database system shall be able to import an ASCII data file,
    convert it as needed, and store the data in the LIMS database man-
    agement system.

6. Interoperability
    The database system shall be ODBC compliant. It will allow data
    exchange with other ANSI SQL, ODBC-compliant database sys-
    tems, including Microsoft Access. Compliance will also enable the
    database to interface with ODBC compliant word processing, statis-
    tical analysis and spreadsheet software for producing reports, letters,
    memoranda and other documents.

7. Data Dictionary
    The data dictionary shall control the definition and manipulation of
    data, and facilitate changes to data structures.

C. **Sample Management and Tracking**

1. Sample Tracking
    Sample tracking shall begin with the sample request and track the
    sample through log-in, analysis scheduling, analysis, quality assur-
    ance, review and approval. An audit trail shall be maintained for
    each sample activity. Sample status will be readily retrieved.

2. Manual Sample Log-in
    A manual sample log-in function shall record data including sample
    collector, sample location, sample date and time, sample type, sam-
    ple receiver, sample received date and time, priority assignment,
    test(s) assigned, and sample splitting and field test data. This data
    shall be posted directly to the database. The log-in function shall be
    flexible enough to provide some degree of user customization, such
    as the addition of custom fields and custom sample identification
    formats, or to define sample types and categories.

3. Multiple Sample Log-in
    A multiple sample log-in function shall be provided. This function
    shall allow a batch of similar samples to be logged in one operation,
    assigning unique sample identifications to each sample, and dupli-
    cating common fields for each sample in the batch. Individual sam-
    ples must then be modifiable at the user's discretion.

4. Auto Log-in
    The LIMS shall be able to automatically log samples according to
    a stored schedule.

5. Data Entry

Data entry functions shall perform immediate database updates. Data shall be available for retrieval immediately after data entry. Historical data from an Access database can be imported into the LIMS database.

6. Sampling Site Information

Static information for sampling sites will be stored in the LIMS. The minimum data elements which will be stored are site id., description, location, type and sample schedule.

7. Electronic Import of Historical Results

The LIMS shall provide the capability to import historical data that is stored in electronic format, particularly ACCESS.

D. Sample Scheduling

1. Routine Samples

The LIMS shall be able to store sample collection locations and the frequency that various routine sample types are to be collected from each location.

2. Automatic Login

The LIMS shall be able to log in routine samples automatically including the following:

- Daily routine samples
- Samples for specified days of the week
- Monthly samples
- Yearly samples

3. Automatic Test Scheduling

For routine automatically logged samples, the LIMS shall be able to master schedule the test/analyses which will be required. The schedule shall include:

- Daily routine samples
- Specified days of the week
- Monthly samples
- Yearly samples
- Quarterly
- Semiannually

4. Sampling Site Information

Static information for sampling sites will be stored in the LIMS.

The minimum data elements that will be stored are site identification, description, location, and sample schedule.

E.   Sample Collection

Barcode Sample Labels: The system shall permit printing sample identification labels with or without bar codes and reading/writing barcode labels style 128.

F.   Sample Identification

1. Unique Sample Identification

The LIMS shall automatically assign unique identification codes to each sample. In the case where a sample is split or subdivided, the LIMS shall assign and associate subsequent identification codes with the original sample.

2. Priorities
The LIMS shall allow user prioritizing of samples and their subsequent subparts and splits.

3. Sample Labels
After uniquely identifying a sample, the LIMS shall be capable of providing labels for affixation to the sample container. The LIMS shall provide a standard format that can be duplicated and modified by an authorized user permitting various types of data to be retrieved from the database and incorporated on the label. The standard label format should include room for mulitple fields besides the bar code and be user configurable. Modifications shall allow including special handling or safety procedures. The system shall provide the ability to specify the number of copies of the labels to generate, and shall provide a reprint option for single or multiple additional labels.

4. Bar Codes
The LIMS shall be able to generate and read bar code style 128 for identification, utilization on labels, chain of custody documents, and data entry purposes.

G.   Sample Receiving

1. Receiving Details
When samples arrive at the laboratory, the LIMS shall capture, at a minimum, the following receiving data items:

*   Date and time of receipt
*   Sample receiver

- Location of sample
- Date and time of sample collection
- Sample collector
- Unusual sample conditions
- Test required (if not previously defined)
- Tests requested
- Field test results
- Comments or ability for custom fields

2. Multiple Entry Methods
   The LIMS shall permit entry of the receiving details in multiple ways:

   - The LIMS shall be able to simultaneously log in and receive samples into the LIMS that are unexpected or nonroutine.
   - Samples of a particular type that arrive in batch shall be received in batch. It shall not be necessary for the user to re-enter similar or repeat information for a series of samples.

3. Storage of Procedures and Tests
   The LIMS shall store information including tests required, lab sample preparation, sample holding time, and/or storage requirements with each sample type, such that the LIMS or the user can associate these tests, procedures and time limits with an incoming sample.

4. Associate Procedures and Tests with Samples
   Upon receipt of a sample, the LIMS shall associate appropriate preparation procedures and tests required for specific sample types. User shall be able to add or delete assigned tests.

5. Test Assignment Modifications
   Authorized users shall be able to modify tests or procedures assigned to logged in samples without modifying the standard procedures and test assignments.

6. Calculate Maximum Holding Time
   Based on sample types and tests required, the LIMS shall associate sample holding times with each sample based on its sampling time to produce maximum holding time/date(s).

H. Test/Analyses Administration

1. Standard Tests/Analyses per Sample Type
   Each test or analysis/type shall be uniquely identified with a code by the LIMS. The test identification code shall permit the association of multiple test components with that test code. The LIMS shall store

data about each component such that the user can indicate, upon
initial entry of the data, which components require computer performed mathematical computations.

2. Associate Developed Calculations with Tests
   In order to perform mathematical computations automatically, the
   LIMS shall permit the development and association of mathematical
   routines developed by authorized users for designated test codes.

3. Test Data Modification
   Modifications and deletions of test data by authorized users shall be
   permitted.

4. Test Result Entry
   Test results shall be entered in multiple formats. The LIMS shall be
   entered in multiple formats. The LIMS shall provide the entry of
   test results in the following formats, at a mimimum:

   - All results from one test performed on many samples.
   - All results from many tests performed on one sample.
   - All results from one test performed on one sample.

5. Special Result Values
   The LIMS shall be able to record special result values such as not
   detected, not measured, <, or null. The LIMS shall have the capability to correctly handle all special result values in mathematical computations. Users shall be able to define in advance how special result
   values will be handled in calculations.

6. User ID
   The LIMS shall be able to identify and capture data concerning
   which laboratory analyst performed the test, and which user entered
   the results

7. Instrument Interface
   The LIMS shall be capable of receiving results directly into its database from interfaced instruments. Specific instruments and required
   processes are listed in section 3.0: Interface Requirements.

I. Bench Sheet/Work Assignment

1. Work Assignment Features
   The LIMS shall provide work assignment features for planning and
   scheduling the laboratory's workload. These features shall take into
   account such data as:

- Sample priority
- Maximum valid holding time
- Sample age
- Due date

2. Work Assignment Reports
   A work assignment report, selectable by the following criteria, shall be provided:

   - Identical analysis type
   - Individual analyst
   - Individual workstation
   - Date

3. Bench Sheets
   The generation of the bench sheet shall be available upon request by a user or in a batch process. Single and/or group selection for reprinting shall be available upon request. The LIMS shall provide the capability to create an additional bench sheet for samples received after the original bench sheets were prepared. The ability to delete a sample or an analysis after it has been scheduled shall also be provided.

4. Bench Sheet Flexibility
   Bench sheet shall be created for one type of test and associate all samples assigned to that test to a bench sheet, as well as a bench sheet for one sample and all assigned tests.

5. Bench Sheet Contents
   Content of the bench sheet shall include, but not be limited to, the following characteristics:

   - Specfic analysis format (e.g., description of analysis, sample name, location, identity, sample date, analysis date, and name of analyst).
   - Quality control samples: blanks, replicates and quality control spikes and standards.

J. Status Monitoring

1. Sample Status
   The LIMS shall provide methods for monitoring sample status throughout the sample life-cycle. Sample status codes shall automatically be assigned and updated by the system based on events or transactions occurring.

2. Test Status
    The LIMS shall provide a method to monitor test and analysis status. The status of tests assigned to a specific sample identification code shall have a direct bearing on the status of the sample itself (e.g., a sample shall not be indicated as complete unless all assigned tests have a status of complete.)

3. Sample Status Codes
    The LIMS shall provide codes to monitor sample status for the following conditions, at a minimum:

    - Sample expected or logged, but not received
    - Sample received by the laboratory
    - Sample has tests assigned that are in progress
    - Sample has all assigned tests completed
    - Sample results have been reviewed and verified
    - Sample data has received formal approval from lab management
    - A recollection of the sample has been ordered
    - Broken sample container
    - Custom status codes defined by the laboratory

4. Test Status Codes
    The LIMS shall provide codes to monitor test and analysis status for the following conditions, at a minimum:

    - Test is assigned to a bench sheet, and is in progress
    - Test is complete and results have been entered into LIMS
    - Test results have been reviewed
    - Test results have failed quality control
    - Test results have exceeded specified limits
    - A retest has been ordered for the same sample and test
    - Test results have associated text or limits violations

5. Sample Disposal

    The LIMS shall provide a means for users to know when samples may or should be disposed of.

K. Test Result Management

1. Comments
    The LIMS shall permit the entry of comments and/or coded comments, which may be inserted by users in place of, or in addition

to, analytical result data. The LIMS shall permit the user, at the user's option, to enter an explanation in textual format to describe unusual conditions or circumstances. When test has been added to explain a test result, the LIMS shall indicate that associated text exists.

2. Calculations

   The system shall support calculations based on the results of muliple analyses and perform reasonableness checks on the computed results. The number of significant digits for calculations shall be user definable.

3. Results Limits

   Test data shall have associated results limits. The LIMS shall allow users to enter regulatory limits such as MDLs and MCLs and associate set of limits with each sampling location. Each analyte in a limit set shall have associated effective dates. These limits shall be used by the LIMS transaction programs to check results being entered and flag the user, during result entry, regarding adherence to the limits.

4. Multiple Limits Sets per Location

   The LIMS shall include the ability to specify multiple sets of limits for each sampling location. Each location shall have an associated primary limit set. All other limit sets at a location shall be considered as secondary limits.

5. Test Result Review

   The LIMS shall allow an authorized user to review test results. The review of test results shall be permitted in multiple fashions; by individual test code, by individual samples and a range of identification code(s), by analytical result date, and by bench sheet.

6. Historical and Precision Level Comparisons

   For assitance in reviewing and approving test results, the LIMS shall allow the user to read historical results for sample locations and analyses. Precision levels of the analytical results based on Quality Control results shall also be available to the user.

7. Review Actions

   The review function shall allow the following actions:

   - Reviewer indicades agreement or disagreement with the test result.
   - Reviewer requires a retest, where a retest is defined as a multiple

of the original performance of the test. The results from a retest shall be associated with the original sample identification and test code.

- Reviewer requests that the sample be collected from the same location again to rerun the test. This new sample will be associated with the original sample even if assigned a new sample number.

8. Review Actions Affect Status
   Actions by the reviewer shall automatically update the status of samples and tests.

L. Data Validation

1. Validation of Data Entry
   The validation of all data, including Quality Control (QC) data, shall be completed by the LIMS immediately after entry, so that warnings and reruns are indicated to the users as soon as possible. The LIMS shall prevent the entry of clearly invalid data in key data entry fields.

2. On-Line Help
   An on-line help facility shall be provided with the LIMS. Help shall be available for each functional portion of the system, such that a user can request help information and then return to their original position upon exiting the help function.

M. Chain of Custody/Audit Trail

1. Chain of Custody Documents
   Chain of Custody (COC) documents shall be produced by the LIMS for each sample bottle collected. In general, the chain of custody may be printed in conjunction with the sample labels. An authorized user shall be able to reprint chain of custody documents on request.

2. COC Appropriate to Sample Type
   The chain of custody documents appearance shall be tailored to the specific sample. Sample identification and bar code, location, sample type, preservatives required, special instructions, and tests requested shall be printed on the chain of custody. The chain of custody document shall include space for the sampler to write in date/time collected, collector's name, field test results, comments, and at least two signature/date lines for transferring sample custody. An authorized user may modify the format and content of the chain of custody document.

3. Audit Trail for Changes
The LIMS shall provide a complete audit trail of data entry and modification to maintain and verify data integrity. Such fields as date, time, old data values, reason for modification, and responsible party shall be recorded when data updates are made.

N. Sample Approval

1. Final Approval
The LIMS shall provie a function for an authorized user to approve all associated sample and test results data in order to complete the chain of custody requirements, and make the data available for use by other departments and in regulatory reports.

2. Multiple Approval Formats
The approval of sample data shall be permitted by individual sample identification code, by test type, by collection location, and by analytical result date.

3. Management Approval or Disapproval
This function shall allow a manager to indicate their approval or disapproval with the sample test result information. The LIMS shall permit the authorized user to disapprove a sample and its associated data when it is discovered that some portion of the data requires a modification after the original approval. This action shall be recorded in the chain of custody audit trail.

4. Protection of Final Management Approved Results
Once the final approval has been completed, LIMS shall provide the ability to prevent any further modifications to the sample and its associated data.

O. Quality Control

1. Sample Results with QC Sets
The LIMS shall provide a means of calculating, storing, and retrieving Quality Assurance (QA) data such as blanks, spikes, duplicates, % recovered and quality control (QC) standards, and shall provide a method of associating sample analysis results with a set of quality control data for specific batches.

2. QC Calculations and Graphical Reports
The LIMS shall include the ability to generate precision and accuracy data by calculating standard deviation from replicate samples and QC standard. The LIMS shall construct and update QC charts using standard deviation, QC standard trending, data validation

through predefined QC criteria, historical concentration ranges, and
regulatory standards. Trending capabilities shall include the tracking
of consistent bias.

P.   Statistical Analysis

1.   Analysis and Graphics
      The LIMS shall include or provide an easy interface to a standard
      product for statistical analysis capability for historical trending and
      examination of LIMS data. Graphics capabilities shall also be pro-
      vided for display and reporting of statistical information.
2.   Graphics
      The graphics component shall be able to produce a variety of charts,
      plots, and maps.
3.   Interface Requirements
      If the statistical analysis and/or graphics functionality are not part
      of the standard LIMS, a seamless interface between a recommended
      product and the LIMS is preferred. If such an interface is not avail-
      able, the Proposer shall detail the procedure that will need to be
      followed by the user to use the statistical or graphical software in
      order to meet this requirement.

Q.   On-Line Queries

1.   Ad Hoc Queries
      End-users shall be able to retrieve logically related data, quickly and
      easily, in an interactive environment, without the need for a detailed
      understanding of data storage and programming techniques.
2.   Multiple Query Criteria
      The LIMS data inquiry facility shall provide efficient retrieval of
      sample data based on sample identification code, location, analyst
      name, date received, workstation or device, test, analyte, result val-
      ues, sample type, and sample status.
3.   Structured Query Language Tools
      End-user tools that use an SQL interface shall be provided. The
      LIMS shall provide the user with a query facility which supports
      nested query, table joins, and outerjoin functionality.
4.   Standard Queries
      The LIMS shall provide standard queries for, at least, a specific sam-
      ple's associated data, all results for a specific sample collection loca-
      tion, status of samples, status of tests being performed, and all ad-
      ministrative or static data.

5. Multiple Output Options
   The query function shall be capable of displaying query results on the user's workstation screen, sending them to a printer or saving them as an ASCII file. Saved queries shall be exportable through, or accessible from, ODBC drivers.

R. **Information Reporting**

1. Report Development
   Company A needs to generate State and Federal regulatory reports, trend analyses, QA/AC charts, and graphically formatted reports for administrative planning purposes. The LIMS shall provide or recommend a third party report development tool that is capable of integrating a wide variety of data types from multiple sources. Information from the LIMS database shall be available for report generation. This reporting tool shall include the following minimal capabilities:

   - ODBC compliant
   - GUI development interface
   - Calculations such as total, subtotal, subtraction, addition, multiplication, division, average, maximum, minimum, standard deviation, mean, median, and mode
   - Format options such a font size and type, page headers and footers, number of significant digits
   - Merging graphics, charts, and text into a single report
   - Retrieve and integrate data from Microscoft Access databases as well as the LIMS database
   - Create barcharts, trend lines, pie charts with retrieved data

2. Preprogrammed Reports
   The following set of preprogrammed LIMS reports shall be provided:

   - Samples received for a user-specified time frame
   - Test results report, including comments
   - Work backlog report by sample status
   - Work backlog report by due date (sample aging)
   - Work backlog report by priority
   - Test results out of limits report
   - Quality control sample report

3. Workload Management Reports
   Workload management reports shall be provied to assist with inter-

pretation for work assignment, staff load balancing and laboratory preference. The following types of reports shall be provided as part of the standard LIMS software:

- Sample volume report (number of samples processed)
- Test volume report (number of tests performed)
- Turnaround time report from sample receipt to approval, summarized by analysis)
- User-definable reports
- Ability to save report to disk for submission to EPA.

3.0 Interface Requirements

A. Electronic Instrument Interface

1. Interface with the following instruments:

   Perkin Elmer Elan 6000 ICP-MS
   Shimadzu TOCC 5000 analyzer
   Perkin Elmer 2100 AA instrument
   Perkin Elmer 3030 AA with WinLab
   Perkin Elmer 8500 GC with Turbochrom 4.1
   Dionix Ion Chromatograph 500 with Peaknet
   Finnegan GCQ GC-MS
   Any instrument with an RS-232 port

2. Provide a method to identify each instrument uniquely.
3. Able to receive and process analytical control sample results from instruments.

B. Transferring Information

1. Unique Device ID
   In order for the LIMS to acquire test results from laboratory instruments, the LIMS shall provide a method to identify each device uniquely.
2. Direct Data Transfer
   The LIMS shall be able to receive and process analytical and quality control sample results directly from instruments that produce final results while the instrument is operational and without disrupting other LIMS users.

3. Data Processing

    After processing or data reduction, the LIMS shall be able to receive
    and process analytical and quality control sample results from PCs.
    The selected vendor shall provide the software required to transfer
    the data to the LIMS.

4.0 Optional LIMS Functionality

The Proposer is requested to respond to items listed in this section only
if they can provide the specific functions, or can interface to third-party
packages that meet the functional requirements. When an interface to a
third-party software is required, the Proposer shall clearly identify the
recommended package(s).

A. Cost Accounting

1. Company A may wish to associate labor and/or material cost with
    specific samples and analysis types. The LIMS shall provide, at a
    minimum, the ability to associate appropriate accounting codes with
    the LIMS data. This function shall provide a means of tracking costs
    for analytical purposes regarding specific projects or cost centers.
2. The Proposer shall describe all accounting features available with
    their LIMS product.
3. This should be an optional feature which can be turned on and off
    as required. It must not be necessary to invoice samples.

5.0 Product Support

A. Technical Support

1. The Proposer shall provide support for all software products in-
    cluded under this contract. Prior to Final Acceptance, the Proposer's
    support staff shall respond within 2–4 hr to all support calls placed
    during normal business hours, 7:00 a.m. to 5:00 p.m. Eastern Stan-
    dard Time, Monday through Friday. Support calls placed after nor-
    mal business hours or on Saturday and Sunday shall be responded
    to within 4 hr on the first regular business day following notification.

2. One year of support shall be provided under this contract (from LIMS System Formal Acceptance date). The support agreement shall be renewable on an annual contract basis.
3. The Proposer shall provide a toll-free telephone number for support calls.
4. The Proposer shall indicate if local and/or national user groups exist for each software product identified in their proposal.
5. Remote diagnostics, bulletin board/Internet support.

B.   Upgrades/Fixes

1. Functional fixes to the software shall be provided as they are released at no extra cost. Supporting documentation for hardware and software reflecting modifications shall be supplied, when necessary, at no extra cost.
2. For as long as Company A maintains an active support agreement, upgrades and enhancements to the software shall be provided automatically at no additional cost. Supporting documentation for software reflecting upgrades and enhancements shall be supplied at no extra cost.

C.   Documentation

1. Company A shall have placed the LIMS source code in escrow.
2. The Proposer shall provide complete hard and soft documentation for the LIMS application and the instrument interfaces. This shall include installation instructions, system administration and maintenance, technical reference and users manuals and any other manuals relevant to the selected LIMS application.
3. A simple step-by-step users manual shall be provided for the end users.

6.0.   Training

A.   LIMS System

1. The selected Proposer shall train the laboratory and systems personnel in the use of all LIMS application software. Initial training shall be conducted on-site at Company A. Follow-up training can be provided on-site or at regional training centers.
2. The selected Proposer shall provide all instructors and instructional

material including trainees' workbooks, instructor guides, training aids, equipment, and technical manuals.

3. The selected Proposer shall coordinate with Company A regarding use of facilities if courses are to be held on-site. Equipment and software that are provided as part of this contract may be utilized for training, provided they are not adversely affected. Any equipment or software modified for training by the Proposer shall be restored to its original condition.

4. Courses that include general programming elements shall provide instruction such that the attending student will be capable of programming related software applications and/or modifications without guidance, or with only minimal supervision. This requirement applies only to the software supplied by the LIMS Proposer.

5. At a minimum, required courses are as follows:

- End-User Training: Provide training sessions on-site that instruct 10 endusers in the overall use and operation of the LIMS application software.
- System Administration Training–Provide training on-site for two (2) owner designated personnel who will act as system administrators for the LIMS computer configuration and applications. The training shall include LIMS administration tasks, software management functions, and computer security. The training shall also include complete system backup and reload procedures, file management utilities, and system generator procedures.

7.0 Installation Services

A. Services

The Proposer shall provide installation and startup services including formatting all disks, loading required software on the LIMS server, client workstations, and instrument PCs, and creating all necessary custom command files to activate the system automatically upon startup.

B. Documentation

Complete hard and soft copy documentation of the LIMS application software and the instrument interfaces shall be provided to the users by the time of installation. This includes users' and reference manuals.

8.0   Functional and Acceptance Testing

A.   Functional Testing

The selected vendor must provide a test plan and perform testing on the system after installation to demonstrate functionality and performance. This will be a checklist that verifies the specific functions and capabilities of the selected LIMS that are required by Company A and detailed in the Technical Specifications of this document.

B.   Acceptance Testing

The acceptance test period runs for the first 120 days after successful completion of the functional testing. During this period, the LIMS will be utilized by the laboratory staff in day-to-day operations. The purpose is to test the LIMS stability and completeness over time. The selected vendor shall provide the following services during the installation and acceptance period:

- Telephone assistance to users in operation of the system.
- Resolution of deficiencies noted during the functional test and acceptance testing period.
- Correction of software failures.
- Upon notification of failure (via telephone call to designated telephone number), diagnose and provide fixes or work-arounds to the failed software. Provide assistance necessary to return the system to correct operation.

C.   Final Acceptance

Final acceptance is accomplished by successful functional testing and successful completion of the 120 day test period as determined by Company A.

IV.   TABLE OF COMPLIANCE TO SPECIFICATIONS

Instructions

1. The Proposer shall complete this Table of Compliance to Specifications and return the completed form as part of their proposal.
2. A complete description of each requirement is in the LIMS Specification portion of this Request for Proposal.

3. For answers of "Comply with Modifications" describe the modifications including cost and time required. Use the following format:

- Header: Modifications Required for Compliance
- Table of Compliance item number
- Specification reference number
- Modification description: Use as many lines as necessary
- Costs
- Time required after award of contract

4. If your product exceeds the minimum requirements, describe how it exceeds. Use the following format:

- Header: Exceeds Minimum Requirement
- Table of Compliance item number
- Description

1. OVERVIEW

1B. SYSTEM CONFIGURATION

	REF.	COMPLY	DO NOT COMPLY	COMPLY WITH MOD.
The LIMS shall be compatible and run on an ethernet network with the Novel Network 4.x network operating system	B1			
The system shall run on the Company A Server as previously described	B2			
The system must be an ACCESS or SQL Based client/server application	B2			
The system's client software will run on 133 MHz Pentiums with 64 MB RAM	B3			

2. LIMS REQUIREMENT

2A. SYSTEM MANAGEMENT

	REF.	COMPLY	DO NOT COMPLY	COMPLY WITH MOD.
The LIMS shall be licensed for 8 concurrent users not counting the interfaced instruments. Up to 10 workstations shall have access to the LIMS	A1			
The LIMS shall run on a server platform and an operating system compatible with a Novell Netware 3.12 and 4.x.	A2			
Provide system management tools as defined in the Scope of Work Technical Specifications	A3			
Provide owner definable security by user, user group, function	A4			
Access levels shall:				
Restrict user of system level functions (such as user authorization)	A4			
Restrict the use of data entry, data approval, data retrieval, data modification, database structure creation or modification functions	A4			
Provide a means to archive data:				
Include collection and approval data ranges, sample type, location, and test	A5			
At the request of system administrator	A5			

Automatically after a specified period of time	A5			
Include user-selectable parameters	A5			
Provide a means to purge data:				
At request of system administrator	A5			
Parameters shall include collection and approval date ranges, sampling point and sampling type	A5			
Automatically after a specified period of time	A5			
Includes user-selectable parameters	A5			
Maintain static administrative or business rules information	A6			
Authorized users shall be able to query, add/modify, and delete this administrative and rule information	A6			

2B. Database Management

	REF.	COMPLY	DO NOT COMPLY	COMPLY WITH MOD.
Provide a relational database management system (either AC-CESSS or SQL) for information storage and retrieval	B1			
The LIMS RDBMS shall be available with full use license, providing not only access to the LIMS application, but also:				

Application development tools	B1			
A data dictionary	B1			
A data query utility	B1			
A report writer	B1			
The RDBMS shall be licensed for 18 users	B1			
Database development tools shall be licensed for two users	B1			
The RDBMS shall support client/server architecture	B1			
The RDBMS shall support parallel processing	B1			
The RDBMS shall support data spanning multiple physical disks	B1			
A transaction journal utility shall provide database reconstruction in case of system failure	B2			
Interactive database management tools shall be a GUI interface	B3			
The RDBMS shall be able to export data into the following formats:	B4			
ASCII	B4			
Excel	B4			
Lotus	B4			
RDBMS shall be able to import an ASCII data file	B5			
Historical data from an ACCESS database can be imported into the LIMS database	B5			
The database shall be ODBC compliant and will allow exchange of data with other ANSI SQL, ODBC-compliant database systems such as MS Access	B6			

The database dictionary shall control the definition and manipulation of data and facilitate changes to data structures.	B7		

2C. Sample Management and Tracking

	REF.	COMPLY	DO NOT COMPLY	COMPLY WITH MOD.
Sample tracking shall track the sample from login, analysis, quality assurance, review and approval. Sample status will be readily retrieved.	C1			
A manual sample log-in function shall record data including				
Sample collector	C2			
Sample collection date/time	C2			
Sample receiver	C2			
Sample received date/time	C2			
Sample location code	C2			
Sample location	C2			
Tests assigned	C2			
Priority assignment	C2			
Field analysis data such as temperature, pH, chlorine residual	C2			
The log-in function shall allow user cutomization:				
The addition of custom fields	C2			
Custom sample identification formats	C2			

Provide a function to login multiple similar samples in one operation. Individual samples must then be able to be modified at the users discretion	C3			
The LIMS shall be able to login samples automatically according to a stored schedule	C4			
Data shall be available for retrieval immediately after data entry	C5			
Data entry functions perform immediate database updates or inserts	C5			
Store static information for sampling sites:				
Site ID	C6			
Description	C6			
Location	C6			
Sample type	C6			
Provide the capability to import historical test result data stored in an ACCESS database	C7			

2D. Sample Scheduling

	REF.	COMPLY	DO NOT COMPLY	COMPLY WITH MOD.
Store locations for routine sample collection	D1			
Be able to login routine samples automatically including the following:				
Daily routine samples	D2			

Samples for specified days of the week	D2			
Monthly samples	D2			
Yearly samples	D2			
For routine automatically logged samples, be able to master schedule the required tests including:				
Daily routine samples	D3			
Specified days of the week	D3			
Monthly samples	D3			
Yearly samples	D3			
Quarterly samples	D3			
Semiannual samples	D3			
Status information for sampling sites shall be stored in the LIMS	D4			

2E. Sample Collection

	REF.	COMPLY	DO NOT COMPLY	COMPLY WITH MOD.
The system shall permit printing sample identification labels with or without bar codes and reading/writing barcode labels style 128.	E1			

2F.   Sample Identification

	REF.	COMPLY	DO NOT COMPLY	COMPLY WITH MOD.
Ability to assign unique identification codes to each sample	F1			
Able to prioritize samples	F2			
Can provide user definable sample labels	F3			
Provide the ability to specify the number of labels needed to allow a reprint option	F3			
Able to generate and read barcode style 128 codes for identification, utilization on labels, chain-of-custody forms	F4			

2G.   Sample Receiving

	REF.	COMPLY	DO NOT COMPLY	COMPLY WITH MOD.
The LIMS shall be able to capture the following information at sample login:				
Sample collector	G1			
Sample collection date/time	G1			
Sample receiver	G1			
Sample received date/time	G1			
Sample location code	G1			
Sample location	G1			

Tests assigned	**G1**			
Priority assignment	**G1**			
Field analysis data such as temperature, pH, chlorine residual	**G1**			
Comments or capability for custom fields	**G1**			
The LIMS shall permit entry of receiving details in multiple ways:				
Simultaneously login and receive unexpected or nonroutine samples	**G2**			
Samples of a particular type that arrive in batch shall be received in batch	**G2**			
Store information with each sample type including:				
Tests required	**G3**			
Lab sample preparation procedures	**G3**			
Holding times	**G3**			
Sample storage/preservation requirements	**G3**			
Ability to add or delete assigned tests	**G4**			
Associate procedures and tests with samples	**G4**			
Authorized users shall be able to modify tests or procedures assigned to samples without modifying the standard procedures and test assignments	**G5**			

Associate sample holding times with each sample based on its sampling time to produce maximum holding time/date(s)	G6		

2H. Test/Analysis Administration

	REF.	COMPLY	DO NOT COMPLY	COMPLY WITH MOD.
Uniquely identify with a code each test or analysis type	H1			
Permit the association of multiple test components with each test identification code	H1			
Store calculation data about each test component	H1			
Permit the development and association of mathematical routines for designated test codes and components	H2			
Permit modification of test data by authorized user with audit trail	H3			
Provide the entry of test results in the following formats:				
All results from one test performed on many samples	H4			
All results from many tests performed on one sample	H4			
All results from one test performed on one sample	H4			
Able to record special result values such as not detected, <, or null	H5			

Correctly handle all special result values in mathematical computations	H5			
Users shall be able to define in advance how special result values are handled in calculations	H5			
Able to identify test analyst	H6			
Ability to identify user who entered results	H6			
Ability to receive results directly into the LIMS database from interfaced equipment	H7			

2I. Bench Sheet/Work Assignment

	REF.	COMPLY	DO NOT COMPLY	COMPLY WITH MOD.
Provide work assignments features for planning and scheduling the laboratory workload which take into account:				
Sample priority	I1			
Maximum holding time/date	I1			
Sample age	I1			
Due date	I1			
Provide work assignment report, selectable by the following criteria:				
Analysis type	I2			
Analyst	I2			
Workstation	I2			
Date	I2			

Able to generate a bench sheet upon request	I3			
Able to reprint single and/or group selection of bench sheets upon request	I3			
Able to create additional bench sheets for samples received after the original bench sheets were prepared	I3			
Able to delete a sample or an analysis after it has been scheduled	I3			
Create bench sheets for one type of test and associate all samples assigned to that test to a bench sheet	I4			
Create bench sheets for one sample and all assigned tests	I4			
Bench sheet contents shall include:				
Description of analysis	I5			
Sample name	I5			
Location	I5			
Identity	I5			
Sample date	I5			
Analysis date	I5			
Name of analyst	I5			
Quality control samples	I5			

2J.   Status Monitoring

	REF.	COMPLY	DO NOT COMPLY	COMPLY WITH MOD.
Provide methods for monitoring sample status throughout the sample life-cycle login:	J1			
Automatic update of sample status based on events or transactions	J1			
Provide a method to monitor test and analysis data	J2			
Provide codes to monitor sample status for the following conditions:				
Sample received by the laboratory				
Samples expected or logged by not received	J3			
Sample has tests assigned that are in progress	J3			
All assigned tests are completed	J3			
Sample results have been reviewed and verified	J3			
A retest has been ordered	J3			
Broken sample container	J3			
Provide codes to monitor test and analysis status for the following conditions:				
Test is complete	J4			
Test results have failed quality control	J4			

Test results exceed specified limits	**J4**			
Test results have associated text or limits violations	**J4**			
Test is assigned to a bench sheet and is in progress	**J4**			
Test results have been reviewed	**J4**			
A retest has been ordered for the same sample and test	**J4**			
Provide a means for informing when a sample may be disposed of	**J5**			

2K. Test Result Management

	REF.	COMPLY	DO NOT COMPLY	COMPLY WITH MOD.
Permit the entry of comments	**K1**			
Permit the user to enter an explanation in textual format to describe unusual conditions or circumstances	**K1**			
Indicate that associated text exists when text has been added to explain a test result.	**K1**			
Support calculations based on the results of multiple analyses and perform reasonableness checks on the computed results for multiple analyzers.	**K2**			
Allow user-definable regulatory imits or other limits with each sampling location	**K3**			

Use result limits to check results and flag the user during result entry regarding adherence to limits.	K3			
Permit multiple sets of limits per sampling location	K4			
Allow an authorized user to review test result	K5			
Permit review of test results based on:				
Individual test code	K5			
Individual and range of sample identification code	K5			
Analytical result date	K5			
Bench sheet	K5			
Precision levels of the analytical results based on quality control results shall be available to the user	K6			
Allow the user to view historical results for sample locations and analyses.	K6			
The review function shall allow the following actions:				
Agreement or disagreement with test result	K7			
Requires a retest	K7			
Retest shall be associated with original sample identification and test code	K7			
Sample to be recollected from the same location and reanalyzed	K7			
Actions by reviewer shall automatically update the status of samples and tests	K8			

Appendix B
Sample Scripted Demonstration

The LIMS will encompass one main laboratory at the XYZ Plant Laboratory and one satellite lab with future expansion to at least four other satellite labs within 2 years.

The scripted demonstration is to allow the members of the LIMS Evaluation Team to evaluate each product for XYZ's needs. The script also guarantees each vendor a level playing field in the evaluation process. The script is designed to test how the LIMS product will handle a typical sample for our laboratory. The sample chosen encompasses many of the quality assurance, quality control, and data review functions performed on data generated by the laboratory. The data to be used are presented in Section II and in Tables 1–4. The demo will follow the itinerary described in Section I.

I. ITINERARY

1. Vendor Introduction
   During the introduction the vendor will introduce each member of their organization in attendance, including their responsibilities with the company and any pertinent background information.
2. Company Overview
   The company overview will provide information regarding the company's size, past and predicted growth, stability, operating units, and customer base.
3. LIMS Product Overview
   This section will provide the LIMS evaluation team with information regarding the history of the LIMS product as well as the philosophies used in its development.

4. Hardware Platform and Recommended Configuration
    This portion of the presentation is to define the hardware platforms
    supported by the vendor and for the vendor to discuss the details
    of their proposed configuration, including any deviations from the
    hardware requirements specified in the RFP. Please be prepared to
    discuss networking protocols and operation over a LAN and WAN.
    Be prepared to discuss the configuration and implementation of a
    multisite LIMS.
5. File Management
    The vendor will discuss the database management system or file
    structures used by the LIMS.
6. Instrument Interfaces
    The vendor will discuss how their software communicates directly
    with instruments. The vendor will be expected to discuss any special
    hardware and software used to interface the LIMS with the instru-
    ments specified in the RFP. Be prepared to discuss how the LIMS
    will handle the dated data received from an instrument that is not
    year 2000 compliant or an instrument that does not provide a date/
    time stamp.
7. Product Support and Training
    The vendor will discuss the types of training required and recom-
    mended, product customization if required, and pre- and postinstal-
    lation support provided by the vendor.
8. Demonstration and Questions

    • Sample log-in and data entry: see Section III Demonstration
       Tasks
    • Generate a benchsheet: see Section III Demonstration Tasks
    • Generate QC charts: see Section III Demonstration Tasks
    • Scheduling of analyses and data approval: see Section III Dem-
       onstration Tasks
    • Generic LIMS Functions: see Section III Demonstration Tasks
    • Report generation: see Section IV Report Preparation
    • Questions: See Section V Questions

II. DEMONSTRATION DATA

The demonstration data are from a real sample processed through the XYZ
main laboratory and the satellite Q laboratory. This sample was selected due
to the broad range of LIMS functions that would be tested by the data. Limits

were selected for a portion of the data for demonstration purposes. The data have been modified to test the LIMS data entry and quality control features. It is expected that all calculations and limits will be loaded prior to the demonstration. The data, although provided ahead of time for preparation purposes, will be entered the day of the demonstration. The vendor will present the setting of calculations and limits as part of the generic presentation of their system.

You may enter any additional data required by your system or that you wish for clarification purposes. However, any additions should be noted and explained during the demonstration. Additional data should be loaded prior to the demonstration.

Vendors are required to set up and save in advance any queries, ASCII import/export routines, or other items that require more than 5 min to generate and test during the demonstration.

Vendors may deviate from the provided script only if it is necessary to process the sample.

A. Collection Schedule and Log-in (See Table 1 and Chain-of-Custody Form)

The vendor is required to demonstrate the scheduling and log-in of the sample illustrated in Table 1. This sample is routinely collected by A. Sampler for microbiological, organic, and inorganic analysis. A. Sampler logs in the sample at the Q laboratory and analyses the sample for the microbiological parameters. The rest of the sample is shipped to the XYZ lab, where it is logged in and analyzed for the organic and inorganic parameters. The vendor is to demonstrate the logging in of this sample utilizing all the information contained on the chain-of-custody form including the field analysis results for pH, temperature, and chlorine residual. Holding times are calculated from the sample date and not the received date.

B. Analyst Work Assignments (See Table 2)

The analyses have been divided among several analysts and labs.

C. Regulatory and Quality Control Procedures (See Table 3)

Only these control parameters are required for the demonstration. Additional parameters may be set at the vendor's discretion to demonstrate the full func-

Table 1 Sample Schedule and Log-In

Sample Location	Sample type	Analysis	Analysis Lab	Bottle No.
123 sample site	Plant effluent	Total coliform	Q	1
		Heterotrophic plate count	Q	1
		TOC	XYZ	2
		DOC	XYZ	2
		UV254	XYZ	3
		SUVA	XYZ	ª
		THM	XYZ	4
		Alkalinity	XYZ	5
		Iron	XYZ	6
		Manganese	XYZ	6
		Sodium	XYZ	6
		Zinc	XYZ	6

ª This parameter is a calculation
Bottle 1: 125 ml sterile bottle, 30 hr holding time.
Bottle 2: two 40 ml vials, 28 days holding time.
Bottle 3: 250 ml brown bottle, 48 hr. holding time.
Bottle 4: two 40 ml vials, 14 day holding time.
Bottle 5: one 1 L plastic bottle, 14 day holding time.
Bottle 6: one 250 ml plastic bottle, 6 months holding time.

tion of their LIMS. Any additional parameters must be noted during the presentation.

D. Data (see Table 4)

III. DEMONSTRATION TASKS

A. Data Entry

On the day of the demonstration, the vendor will present their preconfigured system. The sample log-in and the data in Table 4 will be manually entered into the system as part of the demonstration. The XYZ staff viewing the demonstration are particularly interested in data entry and review, calculated results, and audit trail functions of the LIMS. Any additional configuration, QC, or analytical data the vendor needs to demonstrate their system completely

Table 2 Analyst Work Assignments

Analyst	Test	Supervisor/ Approval
A. Sampler	Total Coliform	B. Approver
	Heterotrophic Plate Count	
L. Tech	TOC	R. Supervisor
	DOC	
	UV254	
	SUVA⁹	
D. Tect	THMᵇ	R. Supervisor
J. Doe	Alkalinity	R. Supervisor
J. Trade	Iron	R. Supervisor
	Manganese	
	Sodium	
	Zinc	

ᵃ This is a calculated parameter. See Table 4 for required calculation.
ᵇ THM includes (analyzed individually): chloroform, bromodichloromethane, dibromochloromethane, and bromoform.

may be added at the vendors discretion but must be noted during the presentation.

B. Generate a Benchsheet

The vendor will demonstrate how their LIMS generates a benchsheet. The benchsheet should be able to be produced either by analyte or parameter and should include the parameter required, sample number and location, sample type and date of sample collection. The format for the benchsheets is at the vendor's discretion.

Table 3 QC and Regulatory Parameters

Analyte	Category	Type	Limits
Iron	Regulatory	MCL	0.30 mg/L
TOC	Quality control	Duplicate	≤10% difference
Alkalinity	Quality control	Spike recovery	90–110%

Table 4 Data

Analyte	Hold Time	Date Analyzed	Rep. 1 (mg/L)	Rep. 2 (mg/L)	Spike Result
Total coliform	30 hrs	3/23/98	0		
Heterotrophic plate count	30 hrs	3/23/98	14	12	
TOC	28 days	4/5/98	2.80	3.15	
UV254	48 hrs	4/2/98	0.105	0.107	
DOC	28 days	4/5/98	2.70	2.65	
SUVA[a]			see calculation		
Total THM[b]	14 days	4/7/98	see calculation		
Chloroform	14 days	4/7/98	0.002		
Bromodichloromethane	14 days	4/7/98	0.000		
Dibromochloromethane	14 days	4/7/98	0.003		
Bromoform	14 days	4/7/98	0.002		
Alkalinity	14 days	4/3/98	30.5		57.0[c]
Iron	180 days	4/6/98	0.89		
Manganese	180 days	4/6/98	0.03		
Sodium	180 days	4/6/98	25.5		
Zinc	180 days	4/6/98	0.35		

[a] average TOC = (Rep. 1 + Rep. 2)/2; average DOC = (Rep. 1 + Rep. 2)/2; average UV254 = (Rep. 1 + Rep. 2)/2.
If average TOC > average DOC, then SUVA = (average UV254 × 100)/Average DOC else SUVA = (Average UV254 × 100)/Average TOC
[b] Total THM = (chloroform + bromodichloromethane + dibromochloromethane + bromoform)
[c] For alkalinity spike: Spike is prepared by spike 75 ml of sample with 25 ml of 100 ppm standard.

C. QC Charts for Precision and Accuracy

Each vendor will demonstrate how their LIMS generates a quality control chart for precision and accuracy. The charts must contain 20 data points, upper and lower warning limits at 2 standard deviations, and upper and lower control limits at 3 standard deviations. The vendor may utilize any data that they desire, provided the resulting chart shows the precision of replicates, the accuracy of standards, and percentage recovery for spiked samples. A hard copy of these charts will be printed as part of the demonstration. The vendor may demonstrate additional charting features as time allows.

D. Scheduling of Analyses and Data Approval

The vendor will demonstrate scheduling of analyses for weekly total coliform analysis and monthly metal analysis. The vendor may expand on this as necessary to provide a complete demonstration. The vendor is to demonstrate the approval of laboratory data based on Table 2.

E. Generic LIMS Functions

The vendor will demonstrate the more generic functions such as set up of data entry, setting data entry and quality control limits, report generation and cusotmization, data queries, statistics and charting functions, and others.

IV. REPORT PREPARATION

A. Standard Report Format

Attached are copies of the standard report formats used by the XYZ Laboratory. The vendor will develop these standard query/reports prior to the demonstration that are similar in appearance.

B. Custom Reports

During the presentation, the vendor will demonstrate the user's ability to customize individual reports from the standard formats.

C. QA/QC Reports

The vendor will develop a basic QA/AC report for the data.

D. Chain of Custody Report

The vendor will develop a chain of custody report for the samples received by the laboratory that should include all the information contained on the attached sample chain of custody form including the unique laboratory sample id numbers and date/time received and who received the sample. The field results for pH, temperature, and chlorine residual do not need to be included.

V. QUESTIONS

Below is a list of questions that must be addressed by each vendor during the presentation of their scenario. The LIMS Evaluation Team members will rate each vendor based on the information provided. The team members may ask follow-up questions to clarify the information provided. The team members may ask questions at their discretion, if the vendor does not address the topic in the presentation. It is recommended that each vendor review the scenario and provided questions in advance to ensure that these topics are covered.

1. Can the audit trail feature produce a report of all activity on a given sample? On a range of samples?
2. Can the instrument data acquisition system handle intermediate results when a single test requires more than one analysis for a final result?
3. Please explain the data archiving facility.
4. How does the system handle samples that are split between labs?
5. Does the system identify and capture data concerning which analyst performed the test and which user entered the results?
6. Are there reports for overdue samples?
7. Does the system provide a report for out-of-limit violations and indicate which limits were violated?
8. Will the system allow modifications to be made after a sample is approved? Can a sample be unapproved?
9. When reviewing a test, can the supervisor/analyst assign a retest or assign the sample for recollection?
10. If an instrument is interfaced with the LIMS, is there a way to void all or parts of an analysis run if the QC data are outside the limits for any sample in the run?
11. For an instrument interfaced with the LIMS, can the analyst review the data prior to commitment to the database?
12. What problems have you encountered with interfacing the following equipment with your system?

    *   Shimatzu TOC 5000
    *   Dionix 500 Ion Chromatograph
    *   Perkin Elmer Elan 6000 ICP-MS
    *   Perkin Elmer GC

13. How does the LIMS handle data that is below the MDL? How are the data used in calculating spike recoveries?
14. Discuss how characters other than numbers are handled.
15. Discuss the procedure for review and approval of data.
16. Does the system have the capability of readjusting the statistical upper and lower limits each time QC data are entered into the database?
17. Discuss how to add an analyte into the system.

LIMS RFP Evaluation Sheet

Vendor <u>Vendor A</u>   Date Received _____

LIMS <u>ALIMS</u>

Reviewer _____   Date Reviewed _____

General Compliance to XYZ LIMS RFP

	Yes	No
Completed Price Schedule		
Completed Table of Compliance		
Completed Questionnaire		

	Max. Pts.	Pts. Awarded	Comments
Evaluation Summary	800	0	
Schedule −5%	40	0	
Price Schedule/Cost −30%	240	0	
Table of Compliance −50%	400	0	
Qualifications/Experience −15%	120	0	

	Max. Pts.	Pts. Awarded	Comments
Schedule −5%	40	0	
Proposal received by March 9, 1998	30		
Can Provide Software in April/May 1998	5		
Can install software/train in May/June 1998	5		

		Max. Pts.	Pts. Awarded	Comments
Price Schedule/Cost −30%		240	0	
Basic LIMS-SQL Server	(under 50K)	120		
Statical and Quality Assurance	(under 5K)	20		
Instrument interfaces	(under 7K)	20		
Sample Prescheduling	(under 5K)	15		
Laboratory Training Database	(under 2K)	10		
Portable Data Entry Terminals	(under 1K)	15		
Barcoding		10		
Installation/Training	(under 6K)	20		
Cost Accounting		10		

Technical Merit of LIMS −50% (See Table of Compliance)	Max. Pts. 400	Pts. Awarded 0	Comments
1B. System Configuration	20	0	
B1. Novell network compatible	5		
B2. Access or SQL based	10		
B3. Client software will run on Pentiums	5		
2A. System Management	20	0	
A1. Minimum of 8 user licenses	3		
A2. Compatible with Novell Netware	2		
A3. Provide system management tools	2		
A4. Restricted access levels	4		
A5. Archive and purge data	5		
A6. Maintain administrative/business rules	4		
2B. Database Management	20	0	
B1. Access or SQL RDMS	4		
B2. Transaction journal	2		
B3. GUI DB management tools	2		
B4. Export data	3		
B5. Import data	3		
B6. ODBC compliant	3		
B7. Database dictionary	3		
2C. Sample Management and Tracking	25	0	
C1. Complete sample tracking	4		
C2. Manual sample log-in functions	4		
C3. Multiple sample log-in functions	3		
C4. Automatic log-in of scheduled samples	3		
C5. Data immediately available	3		
C6. Static information for sampling sites	3		
C7. Import historical data	5		
2D. Sample Scheduling	15	0	
D1. Store routine sample locations	4		
D2. Automatically log-in scheduled samples	3		
D3. Master schedule routine samples	4		
D4. Store static information of sampling sites	4		

	Max. Pts.	Pts. Awarded	Comments
	400	0	
2E. Sample Collection	4	0	
E1. Reading and writing barcodes style 128	4		
2F. Sample Identification	16	0	
F1. Assigns unique identification codes	4		
F2. Able to prioritize samples	4		
F3. User-definable sample lables	4		
F4. Able to read and generate barcodes	4		
2G. Sample Receiving	20	0	
G1. Captures sample log-in information	5		
G2. Can log-in receive unexpected samples	3		
G3. Store information with each sample type	3		
G4. Ability to add or delete assigned tests	3		
G5. Authorized users can modify tests	3		
G6. Associate holding times with samples	3		
2H. Test/Analysis Administration	25	0	
H1. Unique code for each test or analysis	4		
H2. Association of mathematical routines	3		
H3. Modification of data with audit trail	3		
H4. Entry of test restults in multiple ways	3		
H5. Can handle special results	4		
H6. Able to identify test analyst	3		
H7. Receive results from interfaced equipment	5		
2I. Bench Sheet/Work assignment	15	0	
I1. Work assignments	4		
I2. Selectable work assignment reports	4		
I3. Generation of bench sheets	2		
I4. Benchsheets based on test or sample	2		
I5. Contents of benchsheets	3		
2J. Status Monitoring	20	0	
J1. Monitoring of sample through lab	3		
J2. Method to monitor test and analysis data	3		
J3. Codes to monitor sample status	6		
J4. Codes to monitor analysis status	6		
J5. Informs when to dispose of sample	2		

	Max. Pts.	Pts. Awarded	Comments
	400	0	
2K. Test Result Management	25	0	
K1. Entry of comments	2		
K2. Support calculations	3		
K3. Associate limits with tests/samples	4		
K4. Permit multiple sets of limits	4		
K5. Authorized user to review test results	4		
K6. View QC results and historical results	3		
K7. Review function allows various actions	3		
K8. Automatic sample status update	2		
2L. Data Validation	5	0	
L1. Flag data entry errors at data entry	4		
L2. Available on-line help	1		
2M. Chain-of-Custody/Audit Trail	15	0	
M1. Print COC documents for samples	3		
M2. COC contains required fields	6		
M3. COC format or contents may be modified	6		
2N. Sample Approval	15	0	
N1. Authorized user can approve results	4		
N2. Approval by various fields	4		
N3. Permits disapproval after original approval	4		
N4. Prevents modification after final approval	3		
2O. Quality Control and Statistical Analysis	15	0	
O1. Associates samples with quality control	5		
O2. Generates QC charts	10		
2P. Statistical Analysis	5	0	
P1. Provides historical trending	3		
P2. Produces a variety of charts, plots, maps	1		
P3. Seamless interface w/statistical analysis	1		
2Q. On-Line Queries	15	0	
Q1. Easily retrieve data	4		
Q2. Can retrieve data in a variety of ways	3		
Q3. Nested queries, outer joins, and table joins	2		
Q4. Provides standard queries	4		
Q5. Displays data in various ways	2		

	Max. Pts. 400	Pts. Awarded 0	Comments
2R. Information Reporting	20	0	
R1. Has required minimum reporting tools	7		
R2. Provides preprogrammed LIMS reports	7		
R3. Provides workload management reports	6		

3A. Instrument Interface	20	0	
A1. Instrument Interfaces	16		
A2. Method to identify instruments uniquely	2		
A3. Able to receive/process data	2		

3B. Transferring Information	5	0	
B1. Uniquely identify each device	1		
B2. Receive/Process data from instruments	2		
B3. Receive/Process data from PCs	2		

4A. Optimal LIMS Functionality	5	0	
A1. Associate accounting codes with data	2		
A2. Described accounting features	1		
A3. Feature can be turned off	2		

5A. Technical Support	15	0	
A1. Support for all software by vendor	4		
A2. Support provided with contract	2		
A3. Toll-free number for support	3		
A4. Newsletter & user groups	3		
A5. Remote diagnostics	3		

5B. Upgrades/Fixes	5	0	
B1. Software fixes are free	3		
B2. Upgrades are automatic with paid support	2		

5C. Documentation	5	0	
C1. Provide source code	2		
C2. Complete hard documentation	2		
C3. Easy users' manual	1		

	Max. Pts.	Pts. Awarded	Comments
	400	0	
6A. LIMS System Training	10	0	
A1. Training provides for all applications	2		
A2. Provide all training materials	2		
A3. On-site training coordinated w/ NJAWCo	2		
A4. General programming training	2		
A5. Training for end-users and administrators	2		

	Max. Pts.	Pts. Awarded	Comments
7. Installation Services and Documentation	10	0	
A. Complete installation and startup	5		
B. Complete documentation of LIMS	5		

	Max. Pts.	Pts. Awarded	Comments
8. Functional and Acceptance Testing	10	0	
A. Provide a test plan	3		
B. Correct all failures during test period	5		
C. Acceptance period 120 days	2		

	Max. Pts.	Pts. Awarded	Comments
Company Performance, Reliability, Qualifications (See Questionnaire)	120	0	
A. Hardware Compatibility	8	0	
1. Ethernet networks	4		
2. Specify additional equipment	4		

	Max. Pts.	Pts. Awarded	Comments
B. Relational Database Management	8	0	
3. Explained preferred database	6		
4. Described licensing information	2		

	Max. Pts.	Pts. Awarded	Comments
C. Operating System	**8**	**0**	
5. Windows NT appropriate	**8**		

	Max. Pts.	Pts. Awarded	Comments
D. Export Function	8	0	
6. Listed ASCII options	4		
7. Listed native report writer tools	4		

	Max. Pts.	Pts. Awarded	Comments
E. Front-End Development	8	0	
8. Installed SQL development tools	4		
9. Recommended SQL development tools	4		

	Max. Pts.	Pts. Awarded	Comments
	120	0	
F. Report Writer	8	0	
10. Lists report writer tools	4		
11. Lists recommended reporting tools	4		
G. General LIMS Capabilities	8	0	
12. Tracking of samples	8		
H. Statistical Analysis Graphics	8	0	
13. Statistical and graphical functionality	8		
I. LIMS Interface with Other Products	8	0	
14. Interfaces with word processing/spreadsheets	4		
15. Lotus Notes	4		
J. Instrument Data Acquisition	8	0	
16. Describes interface procedures	8		
K. Product Support	8	0	
17. Describes support options	2		
18. User groups for LIMS	1		
19. User groups for RDBMS	1		
20. Priorities for fixes/enhancements	1		
21. New releases	1		
22. Recommendations for training	2		
L. Proposer Information	16	0	
23. Years selling LIMS	2		
24. Address for supporting office	2		
25. LIMS professionals	2		
Research and development personnel	2		
Software support personnel	2		
26. Number of customers w/LIMS	2		
27. Profile of customer base	2		
28. List of customers	2		

LIMS DEMO EVALUATION

Date: _____   Evaluator: _____

Vendor: Vendor A
System: ALIMS

	Maximum Possible Points	Awarded Points
Hardware Platform and Recommended Configuration	10	
File Management	5	
Instrument Interfaces	15	
Product Support	10	
Training	10	
Demonstration & Questions	50	
(Points are totaled from items listed below)		
Sample Log-in and Data Entry	10	
Generate a Benchsheet	5	
Generate QC Charts	5	
Scheduling of Analysis/Data Approval	10	
Generic LIMS Functions	5	
Report Generation	10	
Response to Questions	5	
	Total Possible Points	Awarded Points
	100	

LIMS DEMO EVALUATION

Advantages	Disadvantages

COMMENTS:

Index

Acceptance, 98, 115, 120, 122
Accounting, 26, 90, 97
Accreditation, 55, 65–67
American National Standards Institute (ANSI), 56
American Society for Testing and Materials (*see* ASTM)
Analysis, 50, 85, 86
Approval, 95
Archive, 40, 41, 72, 87
ASTM, 8,12, 38
Audit trail, 34, 35, 59–60, 62, 86

Backup, 41–46, 71–72, 87
  CD-ROM, 42
  DVD, 42
  Tape, 42–46
Barcodes, 27, 28–30
Bench sheets, 88, 96, 102
Billing, 87
Biometric devices, 30–33, 62

Calculations, 16, 25, 95, 105, 109, 110, 122
Calibration, 57, 65, 66
Central Processing Unit (*see* CPU)
CERCLA, 65
Chain of Custody, 35

Clean Air Act, 65
Clean Water Act, 65
Client/Server, 20, 80, 94
Compliance, 94
Comprehensive Environmental Response Compensation and Liability Act (*see* CERCLA)
Computer systems, 77–80
Control Charts, 90, 96
Cost, 99, 101, 103, 104
CPU, 69, 70, 79

Data, 95
  entry, 33, 108–109, 122
  sheets, 88
Database, 94, 101
Demonstration, 19, 94, 101, 104–105
Design, 7, 11, 12, 19, 48, 51, 115, 117–118, 119
Document control, 56, 57
Documentation 7, 12, 14, 19, 103, 122
DOS, 81

Electronic signature, 61–63, 95
Electronic transfer, 59
Enterprise Resource Planning (*see* ERP)

Environmental Protection Agency
     (see EPA)
EPA, 55, 59, 65, 96
Equipment, 57, 58, 63, 65
ERP, 9, 39, 87–88
Evaluation, 11, 12, 94, 105, 122

FDA, 48, 55, 60, 61–63
Federal Insecticide, Fungicide and
     Rodenticide Act (see FIFRA)
FIFRA, 65
File server, 79
Food and Drug Administration
     (see FDA)
Functional requirements, 10–11,
     12, 13, 21, 48, 101

GALP, 34, 38, 59–61, 95, 107–
     108, 113
GCP, 113
GLP, 113
GMP, 63–65, 113
Good Automated Laboratory Prac-
     tices (see GALP)
Good Clinical Practices (see GCP)
Good Manufacturing Practices (see
     GMP)
Graphical User Interface (GUI),
     51–52

Hard drive, 71
Hardware, 19, 69, 71, 94, 101,
     117, 120, 122, 123

Implementation, 11, 12, 19, 22,
     115, 118–119
Inspection, 56, 57
Installation, 49, 115, 121–122
Instrument, 9, 10, 59, 64, 65, 89,
     97, 103, 121
Instrument integration, 23, 33, 36–37

Integration, 9, 10, 11, 12, 20,
     119–121
International Standard Organiza-
     tion (ISO), 55, 56
ISO 25, 58–59, 65
ISO 9000, 56–57
Inventory 24, 39, 97

Laboratory ID number, 85
Licensing, 94, 101
Local Area Network (LAN), 74,
     75, 79
Log books, 88
Login, 23, 24–27, 85, 95, 102
Look-up, 25

Mainframe, 77–78
Management, 57, 113, 115, 116
Memory, 70, 71
Migration, 52

National Environmental Labora-
     tory Accreditation Conference
     (see NELAC)
NELAC, 65–68
Network, 74, 75, 76, 80, 89, 94, 121

Object oriented, 50
Operating system, 69, 80, 117,
     120, 122
Optical drive, 72
     CD-ROM, 72
     DVD, 72
     MO, 72

Password, 30, 60, 62–63, 111
Personnel, 90
Pharmaceutical Manufacturer's As-
     sociation, 47, 48
Plans, 114
Process control, 56

Processor, 69, 70
Programming, 49
Project definition, 9

Quality assurance, 23, 37, 57, 59,
    63, 107
Quality audit, 57
Quality control, 13, 23, 36, 37, 90,
    96, 103, 107
Quality records, 57, 64, 65
Quality system, 56, 57
Questionnaire, 88, 98–99

RAID, 71
RAM, 70–71
Random Access Memory (see RAM)
RCRA, 65
Records, 57, 58, 61, 62, 63, 64,
    65, 66, 67
Redundant Array of Independent
    Disks (see RAID)
References, 104
Reports, 7, 17, 23, 27, 34, 38, 67,
    86, 87, 89, 97, 105, 110, 122
Request for Proposal (see RFP)
Requirements, 114, 115, 116–117,
    118
Resource Conservation and Recov-
    ery Act (see RCRA)
RFP, 93–99, 101, 114, 122
RS232, 73, 74

Safe Drinking Water Act (see
    SDWA)
Sample collection, 85
Sample identification, 64, 66
Scanner, 29
Schedule, 85, 94
Scheduling, 23, 35, 102
SDWA, 65
Security, 32, 59, 64, 111

Shared processing, 77–79
Single processing, 77
Software, 7
SOP, 59, 60, 90, 107
Source code, 52, 118–119, 123
Standard Operating Procedures
    (see SOP)
Statistics, 96, 103
Status, 96, 102
Support, 7, 97–98, 103, 115, 122
System help, 52

Tape drive, 71–72
Technical requirements, 58
Test, 49, 95, 102
Test sheets, 96
Testing, 114–121
Timing, 113
Toxic Substances Control Act (see
    TSCA)
Traceability, 56
Tracking, 23, 27, 67, 95, 102
Training, 40, 57, 98, 104, 107
TSCA, 65
21 CFR part 11, 30, 61–63
21 CFR parts 211–226, 63–64
21 CFR part s600–680, 63

Unix, 81

Validation, 13, 49, 61, 104, 108,
    113–123
Vendor, 21

Waterfall, 47
Web, 39
Wide Area Network (WAN), 74,
    75
Windows, 81
Workflow, 83–85

T - #0529 - 101024 - C0 - 234/156/13 - PB - 9780824705213 - Gloss Lamination